THE RATIONAL UNIVERSE
EINSTEIN'S BEST IDEA

How Einstein's concept of a "Cosmic Constant,"
derided for decades by critics,
may now control the entire universe.

Prologue:

As part of his 1917 extended theory on gravity, Albert Einstein proposed a force he called the "cosmic constant." This force he added as a computational variable to keep the universe in balance.

He realized the universe needed something more than gravity to keep it from expanding terribly or squashing to nothing. Had Einstein's cosmic constant existed, its force would have extended throughout the known universe and would have exerted its powers evenly.

Not soon after Einstein's revelation, astronomers realized the universe was expanding rapidly; it was far, far larger than people had thought. The idea of a cosmic constant began to seem far-fetched. Such a force would necessarily have been monstrous in its powers while stretching to infinity. This force would have made gravity rather insignificant.

Equally distressing to the cosmic constant concept, scientist made strides to mathematically conquer the smallest atoms. They found the atom to be governed by an extraordinary strong force and a weak force. These forces seemed to have little in common with Einstein's gravity model, and worse, they left no room for something like a cosmic constant.

The double-slit experiment in 1808 had proven that photons must actually exist in ways human brains could not possibly understand. Physicists agreed, quantum phenomena could only be dealt with using mathematical models. And so quantum mechanics evolved.

Visual, rational models had to be abandoned completely.

Einstein eventually gave in. Great scientists like Bohr convinced him that quantum mechanics, though less graceful than a "cosmic constant" vision, would eventually explain everything.

After years of failure and ostracism, Einstein wrote the "cosmic constant" was his "worst mistake." But till the end of his life, he still yearned for a more rational universe.

The concept was forgotten.

But lately, scientific attitudes have changed. Physicists now recognize a vast amount of invisible matter is necessary to keep the universe together and to explain other phenomena--such as why the edges of galaxies speed far too fast to stay attached, but do so anyway.

And current theories now suggest a vast range of phantom-particles—these weird particles blink in and out of existence from empty space, from other dimensions perhaps, to explain observed interactions between elemental particles.

The point is: If modern science can accept such impossibly strange additions to its already weird quantum universe, Einstein's "cosmic constant" seems strangely possible! In fact, let us assume it exists. What is it? And how does it shape the universe it controls?

INDEX

CHAPTER 1
THE RATIONAL UNIVERSE

THE COSMIC CONSTANT

If Einstein's "cosmic constant" exists, it is an incredible storm of black, invisible energy moving forcefully in *all* directions. Since this energy must be in the form of photons—no other elemental energy seems to *exist*—let us call this field the photon storm.

THE PHOTON STORM

Imagine a massive field of photon energy, more intense than exists in some suns—and this invisible black energy moves through us and surrounds us completely everywhere. Why don't we burn up in this intense energy? Why don't we see or feel anything? Well, the answer must be that most of this energy moves right *through* us without consequence.

Science has already confirmed that trillions of neutrinos move through us each instant without affect. So, a tiny photon would be even smaller than a neutrino and would be nearly impossible to detect.

Imagine then an intense storm of energy is flowing through us at this instant evenly in all directions just as Einstein predicted. And we can't even see it is present. It is invisible. And we can't feel it at all.

Or perhaps, we *do* feel it! We feel it a *great* deal. We simply can't identity the force behind our many sensations. If the science is correct, this force creates mass, gravity, stars, and planets—in fact, it creates *everything* we feel. The photon storm creates you, and me, and the entire universe!

Here is how it does it.

The Photon Storm —
An intense storm of photons moving in _every_ direction.

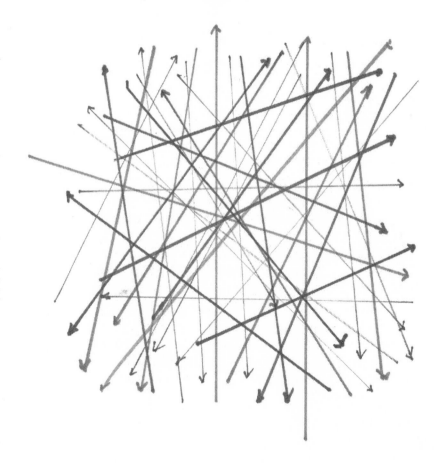

THE PHOTON STORM -- An intense storm of photons moving in *every* direction.

FIRST, SOME QUESTIONS:

What is the photon storm? (Aka cosmic constant? Black energy? Higgs?)

The photon storm is a mass of tiny photons streaming powerfully in *every* direction with extraordinary intensity throughout the universe. Call them god-photons because they encompass everything and are everywhere.

What is a God-photon?

A God-photon is the smallest *possible* amount of moving photon energy. These photons are called god-photons, because, meaning no disrespect, they are *everywhere* present and possess extraordinary intensity. Like Einstein's *cosmic constant*, they exert their powers *evenly* throughout the entire universe.

For the sake of illustration we will view these photon units as actual distinct particles. We recognize quantum mechanics treats elemental particles as clouds of unresolved probabilities. But visualizing unresolved probabilities is impossible, so utilizing actual particle shapes makes the explanations much easier; this process does not otherwise disturb quantum mechanical parameters. The math will turn out to be identical.

IS THIS STORM CONTAINED?

We strongly suspect this storm is self-contained in a finite universe. The reasons become clearer as the configurations unfold.

HOW POWERFUL IS THE STORM?

The present figure is $(1.493)(10)28^{th}$ kg/Cubic meter. This is the (S) value, or STORM value, and it is not a small number. The number has 28 zeroes in it. Imagine this force moving through all of us.

14,930,000,000,000,000,000,000,000,000,000 kg/cubic meter.

This terrific force moves in *all* directions, and it definitely has enough power to do some damage. The (S) force can push matter together or pull it apart.

We will demonstrate in the math section how this (S) figure was derived.

WHAT *IS* A GOD-PHOTON?

The god-photon is a photon like any other—like light or x-rays-- but a g-photon is *very* thin and low energy. It may have a frequency of one and the energy of one plank or less. Because its diameter is nearly *zero*, it can move together in *massive* numbers. It has zero mass, and it moves with zero friction in *all* directions. Like other photons, its wave movement forward actually *assists* the movement of other photons forward in the opposite direction.

We will discuss the derivation of the God-photon later, how it might have come to exist, and why it moves, but for now, just accept that the God-photon— or something quite like it—may exist.

Alternately, we may view the god-photon as a quantum cloud—not an actual particle—but a cloud of probabilities. Assuming these god-photon-clouds have the same characteristics as real god-photons, the mathematical results will be the same.

Why does a God-photon affect only tiny atomic particles?

The God-photon has a diameter nearly zero. It can only react with something substantial, like an atom that it meets *directly head-on.* Space between atoms is immense for god-photons, so most God-photons pass through atoms and entire planets completely without hitting anything, though a few are active around atoms as Fuzz. We will explain fuzz in a moment.

Regular large photons like those in the light and radio ranges react predictably with electrons. These photons are too large to avoid contact. Picture the god-photon as a thin, wiggly wire, with an affective diameter of near zero. Regular photons, on the contrary, are huge ropes with many, many strands. Thus we know large photons exist, and we can utilize such photons for multiple purposes, cooking, vision, electronics; every different wavelength seems to have a different function. If You exist God, thank you very much for large photons.

Unlike large, combined photons, God-photons are too small to see or easily utilize, but, if they exist, their products are everywhere--in fact, they create everything! But we digress--if God-photons exist, they are literally as tiny as energy can get, but they can march together like tiny soldiers in such fantastic numbers, their force flows everywhere with *extraordinarily* power.

How can these photons move in all directions evenly without interference as Einstein wished?

Wave assistance. Now this is where the skeptics begin laughing like hyenas. We welcome skepticism, but just think about it a bit before tossing this book at your neighbor's cat.

The reason God-photons can move so easily in such great numbers is that they push each other forward going the opposite direction. Yes, photons actually possess front to back spin or wave motion. And when any two photons meet they exchange a backwards thrust to conveniently assist each other forward, and— did we mention—this forward speed is the speed of light? Since the universe is *completely* filled with God-photons, they always have something off which they can push--each other.

BUT DON'T FORGET THE QUANTUM CONCEPT; IT CAN ACTUALLY TELL THE SAME STORY.

Alternately, if we view a god-photon as a unique quantum probability cloud (and not a real particle), we simply imagine these clouds constantly exchange energy to create quantum photon-like movements in all directions in the same way as our proposed particles. Visually, we have no idea what goes on inside these clouds or how they physically interact. Though we can measure and mathematically predict certain effects in a quantum universe, we have no visual or mechanical idea how these effects arise.

In both cases—quantum cloud or real particle—the mathematics conjures a universe totally united by exchanged (real or possible) energy-movements. **The concepts do not overtly contradict.**

Imagining the god-photon as a real particle, however, makes the visualizations much easier. And we will follow this process *for illustrative purposes*. In the following drawing, note how the surfers can push each other forward by exchanging a *backwards* thrust. Such assistance would be helpful if the surfers were in empty space and had nothing but each other to push off of.

Imagine now, the surfers are photons, and they fill up space completely. What a party!

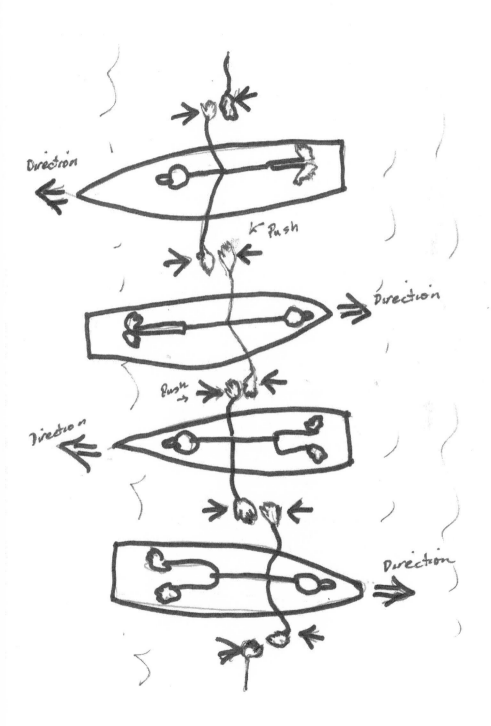

No vacuums exist in the *entire* photon storm universe. Unlike in the classic universe in which photons theoretically move through empty space, the photon storm universe is totally full, totally. In fact, if a photon storm photon were to enter a vacuum, it would not be able to move. How's that? No vacuums! No empty space. The universe is *totally* filled with tiny photons moving shoulder-to-shoulder in *all* directions. Einstein's universe is not empty; space *is* energy.

Now skeptics, hang on. If you laugh too much, you may hurt yourselves. Think about it a bit more. So many things in the classic model are quite difficult (impossible) to conceptualize too, and people have grown forgiving. We ask a little patience here. Photons are weightless, so a little push is all they need. And in this elemental universe, distance-forward is only an abstract, *relative* concept. Light speed, without any opposition, can be whatever God decides—as long as it is a *shared*, relative concept.

Picture a long, thin spring with a diameter approaching absolute zero; it possesses long wave motions pushing it forward, and gains added assistance moving forward from every other photon it meets.

To tell the story again, imagine you are a long wiggling worm, named photon. You wear a baseball cap and are very cute. You are in empty space. You wiggle and wiggle, but you go nowhere! In empty space your wiggling does nothing. Then you find yourself surrounded by trillions of fellow photons. You feel them around you. Now when you wiggle, you push against your friends and you move forward! It's a dance party! In fact, everyone can move forward, and since you wiggle rather hard, you can *all* move forward at the speed of light! Why not?

Thank you very much. What a life! Everything zips forward at the speed of light. We photons also adapt clockwise and counterclockwise *spins*; these spins help us slip easily by each other with the *least* interference. If we touch the side of another photon, our spins always tells us a coordinated way to roll over each other without bother. We create a wonderful tapestry. Call us a photon storm or a happily united photon mesh.

Happy eternities pass, we move across the universe, until, one day, we momentarily run splat into an atom named Bill.

"Ouch! Hi Bill."

We happy photons are now creating mass!

SPIN/WAVE ASSISTANCE FORWARD

God-photons (real particles or quantum probabilities) act to push each other in the opposite direction at the speed of light.

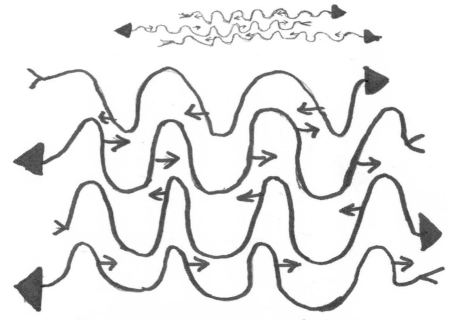

Photons push each other
in <u>opposite</u> Directions.

CHAPTER 2 -- MASS

How does the photon storm create mass?
Mass is not a constant; it is a reaction.

Hold out your finger. Imagine in your finger somewhere is an atom named Bill. Bill is round with two thin legs. He is smiling. Then suddenly, Bill is pummeled by thousands of energetic god-photons from one side. We might think Bill is going to fall over, but simultaneously he is struck by an equal amount of god-photons from the opposite side. In fact Bill is struck from *every* direction simultaneously. Absent an external force to help him, Bill is not going anywhere. Bill can't move.

He continues to stand straight and quite round. Why is he round? Well, he is being pummeled from *all* sides by tiny God-photons. Round is his only choice. And the pummeling goes on forever, never, never stopping ever, not for an instant.

Now Bill *might* resent this constant pummeling, but Bill is philosophical. He realizes the pummeling from the photon-storm is what actually keeps him together. Without the pummeling, his conjoined parts would spin apart completely, and he rather enjoys being together with his friends. So do the trillions of other friendly atoms making up your finger. And so, we hope, do you.

Bill is also advised, most of the photons shooting at him, move through him without hitting anything. Although Bill is more solid than you or me, he is still mostly permeable. He only reflects a tiny, tiny portion of the photon power that comes his way. Different particles can reflect differently. We will discuss reflectivity (Rf) later.

Wave at Bill. He just waved at you. Good.

INERTIA/MASS

So now we have an idea of what creates mass. Bill, an atomic particle, and all his friends in your finger are being pummeled from all sides. This pummeling (or energy interchange) gives your entire finger inertia. In other words the storm tries to keep the atoms in your finger in one place by pushing evenly from *all* directions. If you wish to *move* Bill and your finger, you must supply some sort of external force, and this force is also called acceleration.

You can overcome Bill's inertia by using the force of your finger muscles. The harder you flex your finger muscles, the faster your finger—and Bill—will move. Acceleration times Mass times time equals speed. Again, mass/inertia is provided by the storm of photons pummeling Bill from all directions.

The heftier Bill grows, the more mass/inertia he will acquire and the more muscle energy needed to move him. If gravity pushes on Bill he will also acquire *weight*.

We will get into gravity later, but we can tell you this now. We weren't totally accurate about the photon storm hitting Bill *equally* from all sides. If you look below your finger, you will see something big, really big, even bigger, the earth.

The earth below you is rather large, and so it reflects more of the storm than the sky overhead. This means, fewer god-photons are hitting Bill from the earth side of his body than from the sky side. So Bill is getting hit slightly harder from the sky side; he is being pushed to the earth, and this push is called gravity. Yes, it is a push, *not* a pull. And gravity is pushing on *all* your atoms, just like it is pushing on Bill, giving you what most people wish they had less of, weight.

But wait, forget weight or gravity, we are still talking about mass. The photon-storm moves through our bodies with just a tiny, tiny bit of reaction; God-photons react only with the tiny, tiny atoms in us. By pushing and pulling these tiny atoms from all sides the storm gives these atoms shape, inertia, and keeps them from flying apart.   We are happy it does this, since we like our fingers.   To move our fingers, we require an accelerating force that is applied by our muscles.

Gravity applies a tiny push on all of us from above towards the earth. We don't mind. We know we need gravity too. We don't want to be flying apart. That might not feel good.

All these forces are all beautifully connected mathematically. But to avoid losing the flow we will stick with Bill and pictures, and do the math *later*.

The math is easy. We have Newton and Einstein to help us. They do good work.

Ch. 3 -- GRAVITY
How does the photon storm create gravity?

Well, we already mentioned gravity, but the concept could use some detailing. Gravity is caused by the photon-storm pushing on our atoms. Large objects, and small ones like Bill, soak up or bounce a tiny amount of the photon storm's numbers; so large objects have a photon shadow around them in all directions that has slightly fewer outgoing photons than incoming. The bigger the object, the greater the ratio between incoming and outgoing god-photons and the more gravity that object creates.

So *any* object with mass, bounces god-photons, and so all objects with mass create gravity shadows between them in which the intensity of the storm is decreased. Objects that, for any reason, do not interfere with the photon storm *do not* possess *any* gravity. A particle that does not interfere with the storm also possesses no mass.

MASSLESS

In particular, photons *do not* seem to interfere with other photons in the photon storm. They do the opposite by *assisting* each other's movement. We will explore the exact association later.

So photons, no matter what their size, apparently do *not* possess mass, none. They can't possess mass, because they don't interfere with the storm in a way to give themselves inertia. And so, while huge, very energetic photons may possess *no* mass, at the same time it is possible that tiny, tiny, god-photons, in the right *configuration*, may *acquire* mass, if their configuration causes them to interfere with their fellows in the storm. We will *see* this later.

Gravity shadows in the photon-storm universe are also described beautifully by math. Newton and Einstein have done excellent jobs with gravity already. We will review them later. Hint: We conclude they are right. But their models may need tinkering.

For now, just accept that gravity represents a tiny, tiny, tiny bit of the power of the photon-storm. Most God-photons pass through you and me and entire planets with ease, as if nothing is there. The space in a single atom—from the tiny nucleus to the outer orbitals--is comparable to the space in our entire solar system. Many photons must be bounced or be absorbed to create a *small* amount of gravity.

Gravity is a potent force in our universe, but gravity is created by and totally dependent on the far *greater* power of the photon storm aka (cosmic constant).

In the following picture, with two people walking through space, and with rain coming from *all* directions, the people will be getting wet from their sides, but they will be getting a little *less* wet from the top and bottom because they are holding umbrellas. They will create a shadow for each other. This *small* force on their umbrellas will act to push the people slightly *towards* each other.

Two planets create a similar situation in the photon storm.

GRAVITY,

TWO OBJECTS SHIELD EACH OTHER
AND ARE PUSHED TOGETHER BY PHOTON STORM

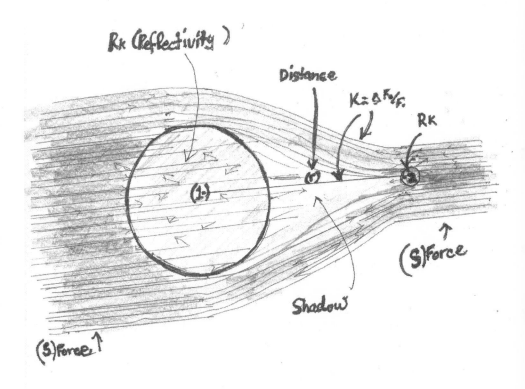

Each planet reflects a portion of the photon storm creating a gravity shadow between them. They are pushed to each other along the gravity shadows between them; these shadows have less intensity than the storm behind them.

Ch. 4 MATTER

How does the photon storm create matter? The electron?

Electrons are apparently quite stable. They can move through space for eons without changing drastically, and they can fulfill connective tasks in minerals without changing apparently for millions of years. Electrons are very conservative chaps. But how can they be so stable? The only thing that might create electrons is the photons in the photon storm, and this storm is moving chaotically all directions.

Well--wait to laugh out loud until we explain the whole concept—perhaps *something* occurs to stick photons together. We don't really know what this something is, but we know, something happens to create a stable particle, something *must* happen or we wouldn't have a universe.

But we have a very good guess. And *if* this guess is correct, it makes the progression very, very, very simple.

POINT STABILITY

What if two elemental photons meet at the *exact* same point, or three photons? Or even four? They would all be rolling about the same exact point. This point would necessarily be very, very small, infinitely small. But all the photons in the group would be united around it. And this entwined union *could* be *very* stable. How stable? Well, what sort of force could unravel this union? The power of the photon-storm is pushing these photons together to the same exact point. No force in the universe is apparently more powerful than the photon storm, particularly at the atomic level. As long as the storm keeps blowing, the union of tiny photons, then, might be *very* stable.

And as a prelude to the electron, we can assume more elemental particles exist, call them pre-electrons that have configurations that are forming but not yet formed. These would be *neutrinos*, one or two or three wrapped God-photons seeking, but not yet finding, a point of total stability.

Such tiny configurations should exist in *massive* numbers, since they are the closest relative to the God-photon, and they would pass through everything without contact, unless, they hit something face-on directly. In fact, such a collision might be enough to align a neutrino's points and convert it to a more stable electron. Thank you very much.

But if the point stability concept seems too incredulous, don't sweat it for now. We will later speculate on alternative configurations and how such stuff is created. For now, just accept that tiny photons are tied together in *some* stable way, and this *stability* makes such configurations *different* than the storm around them. Life in the fast lane is over; such elemental particles develop inertia, mass, spin, and order a credit card.

Again, we are assuming god-photons exist as real particles and not as quantum probability clouds. The math would be the same, since quantum clouds can do whatever particles can do—we simply could not illustrate it. In both cases, however, we can imagine the unity via shared energy of all things.

a) Random Photons

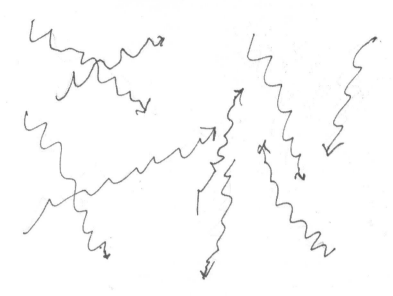

b) Photons Meet at exact point, become stable ELECTRON.

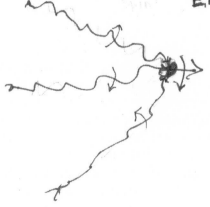

Ch. 5

ELECTRON ATTRIBUTES

FUZZ

In addition, this tiny union might attract hangers-on. We can call these hangers-on, fuzz, Bourne-fuzz some have joked. This Bourne-fuzz acts to add mass to the tiny stable configuration in the same way water molecules attach to a snowflake.

Bourne-fuzz answers most of the PS universe's questions on mass. A configuration can begin rather tiny and nearly massless and immediately attract additional fuzz. Think of Bourne-fuzz as lots of lonely god-photons that attach to an electron like atoms attach to a crystal. Photons are excellent at avoiding other photons, but they get entrapped in the configuration of a slower electron configuration.

So, huge powerful photon bundles, like gamma rays, can still be massless, while tiny photon configurations that interfere with the storm must contain mass or inertia. If Bourne-fuzz does not exist, then the question remains: Why do things possess the mass they seem to possess? We will later discuss alternatives, but for now, we will assume Bourne-fuzz exists, so in a photon storm universe, mass is not an unanswerable question as it is in the quantum universe. Fuzz is it!

Assuming fuzz exists, the math is easy. Each configuration has its own fuzz constant. The fuzz constant describes how many photons align to a configuration at each construction cycle. The math can wait, but most of you with math inclinations, can figure it out on your own. ENERGY = (#) (Fc) (Configuration Level).

Pictures now; math later. (See pic)

c)

Spinning Electron attracts Fuzz

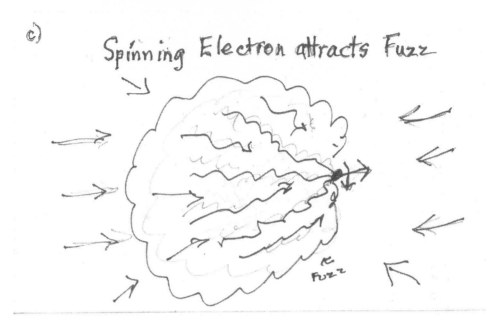

Fuzz

d) Spinning Electrons create Photon Funnels
by spinning storm as it moves through
in both directions.

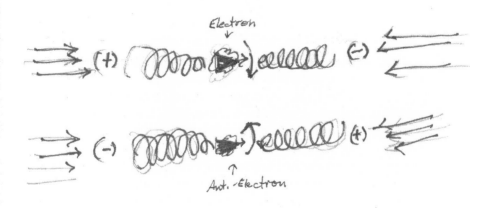

THE HIGGS

Wait a minute! We though stuff got its mass from interaction with the recently discovered Higgs particle! Ain't that right?

In the PS universe the Higgs is a strange kind of phenomenon, call it a ghost-fuzz particle, created momentarily during particle annihilations. When a configuration is blown apart, the fuzz-cloud associated with that configuration suddenly has nowhere to hang. It appears, very briefly, then goes its way.

In the photon storm universe, the Higgs, and other similar apparitions, are happy fuzz-ghosts. We will meet more of these quirky, unstable semi-particles later.

And, in contrast to the classic explanation—*these particles spring from empty space*—we can easily visualize where these particles physically originate, the photon storm (not empty at all).

SPIN (Back to the Electron)

Since the electron configuration is created by photons, and photons possess clockwise and counterclockwise spins, then the electrons must also have spin, clockwise or counterclockwise. In addition, all the photons in an electron bundle—we aren't positive the precise number, though we will later compute a figure—would be chasing the combining point in the same direction, clockwise or counterclockwise. So, electrons have stability from their configuration, mass from Bourne-fuzz, or Higgs-ghosts, and spin because they all spin around an infinitely small point.

ANTI-ELECTRONS

In addition, since electrons spin in clockwise and counterclockwise directions, they must create two *different* varieties of electron, and if these opposite turning electrons run into each other they might unravel. Oops! Their Bourne-fuzz would fly off as energy in all directions as the two particles mutually annihilate. We will call one electron an electron, and the other electron an anti-electron.

ELECTRON DIRECTIONALITY

An electron also has *directionality*, because all the photons are still moving forward in a vector direction. Perhaps the electron does not move at the speed of light—it should be a *bit* slower since it possesses inertia: but it still moves directionally like a photon, attracting fuzz mass and inertia from the storm as it goes.

Electrons and anti-electrons have opposite spins in line with their directionality. For clarity, we will assume a *negative electron spins clockwise* and a *positive anti-electron spins counterclockwise* in line with their directional movements.

Note: Conveniently in the PS model, if an electron gains energy, this energy is in the form of god-photons. These photons push the electron forward, so its velocity increases, as does its mass. These increases match exactly with Einstein's calculations--mass increases as the electron approaches the speed of light.

However, in the PS model the energy is not mystically invisible; the energy is *countable* as actual increases in the *number* of god-photons. This makes the visualization easy. We will do the math later.

CHARGE!!

Finally, an electron possesses *charge*. Charge is produced by an electron's spin twisting the photon storm. Anyone from the US Midwest knows how powerful funnel clouds can become. A spinning electron twists storm photons into funnels in both directions, creating powerful vortexes like tornadoes.

Since photons pull each other forward going the opposite direction, then a negative funnel would suck hard on a positive funnel, and the attraction would be a pull. Two negative or two positive funnels would have the opposite affect, seeking to move *away* from each other.

Electrons produce positive and negative charges, but the charge in line with the electron's directionality is the *defining* attribute, since the electron will move in this charge's direction.

So a regular electron moves directionally toward its *negative* charge and an anti-electron moves toward its *positive* charge. If a positive electron meets a negative electron head-on, they might well unravel each other, splitting up into massless photon energy.

However, if electrons and anti-electrons meet at an appropriate angle, their spins *and* charges could align *perfectly* to create a different *stable* configuration, quarks.

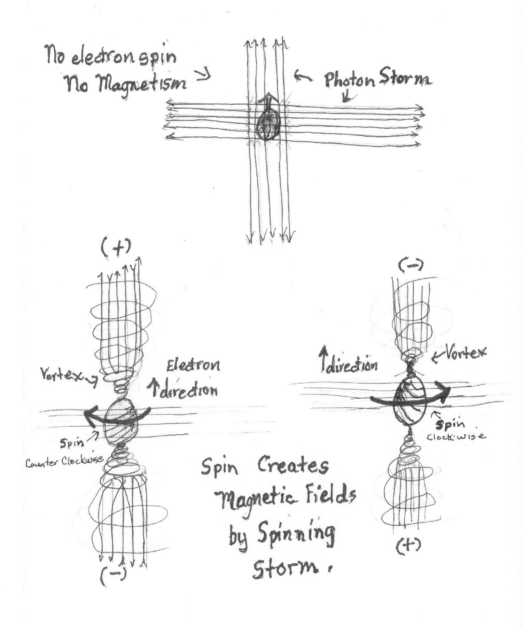

No electron spin
No Magnetism →

← Photon Storm

(+)

Vortex →

Electron
↑direction

Spin →
Counter Clockwise

(−)

↑direction

← Vortex

Spin
Clockwise

(+)

Spin Creates
Magnetic Fields
by Spinning
Storm.

Very much in *opposition* to the mystical situation in the quantum classic universe, the *electron*, electron *charge*, and *electromagnetism* are all very *easy* to *picture* visually in the PS universe. Such pictures should help students understand the *why* of such phenomena.

Ch. 6

ELECTROMAGNETISM

The funnel charges created by electron spin can be magnified exponentially if the electron is in an electric current magnetic field, such as in metal wires. Picture the spinning charges being stretched to tornadoes, each node of the electron pulled in the opposite direction. Spinning fuzz from the storm fills in the gaps.

If the potential acting on an electron suddenly diminishes, the spinning cones of photons will be released as the electron returns to normal. These cones of spinning photons move away with energy equal to the energy absorbed from the current.

In other words, the electrons release electromagnetic radiation in twisted photon bundles in direct relation to the changing current running in the wire.

Such radiation allows us our television, smartphones, and online shopping—shopping without which humanity could not *possibly* survive.

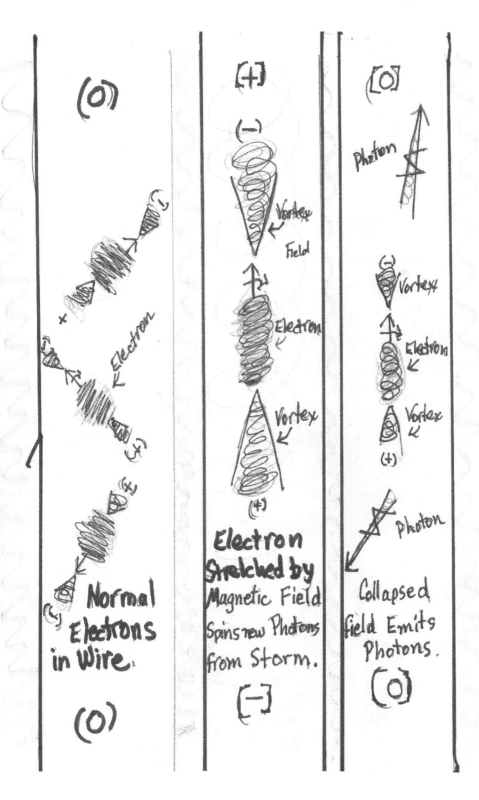

(O)

Normal
Electrons
in Wire.

(O)

[+]

(−)

Vortex
Field

Electron

Vortex

(+)

Electron
Stretched by
Magnetic Field
Spins new Photons
from Storm.

[−]

[O]

Photon

(−) Vortex

Electron

Vortex

(+)

Photon

Collapsed
field Emits
Photons.

(O)

We will speculate further on electromagnetism, but note in the following picture how well the photons released from electrons resemble the combo-photon design, the *simplest* photon design. If this simple design is correct, then large photons might well simply be bundles of elemental photons twisted together like threads.

These threads might well resemble graceful helixes as pictured on the next page. Einstein favored a graceful universe.

THE SIMPLEST PHOTON CREATION (THE THREAD)

To repeat, single tiny, god-photon move every direction without much opposition—they are too tiny to interact with most matter. However, if a spinning electron can catch god-photons in its spinning wheel, these small photons can be woven neatly into a threaded-bundle. This bundle has increasing power according to the number of photons spun together.

This bundle (thread) of photons is large enough to interact with matter and these bundles can be controlled and used in electromagnetic human technologies.

These bundles, photons, once formed, can speed across the universe without changing, until they meet matter of some sort. Matter can absorb these light-bundles, or reflect them, or be permeable to them.

Since these bundles do not clash with the photon-storm around them, they are not matter, and they can move at the speed of light without opposition.

Combo Polarized Photon Helixes

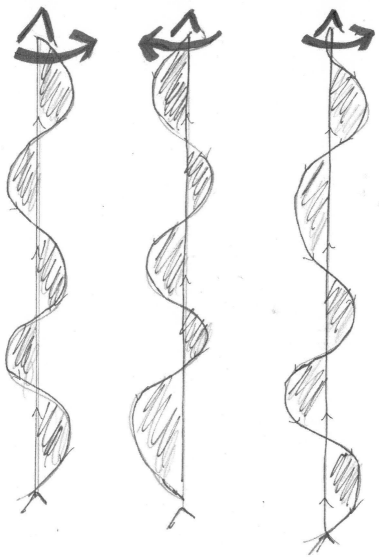

Each god-photon may have the energy of one plank, and/or additions to photon *bundles* may only be possible in specific quantities, each one equal to a single plank. So the measured frequency and total energy of *photon bundles* would be directly related to the *number* of god-photons in that energy bundle.

If the reader is confused on the meaning of Planck's constant, the concept is explained more directly in the math section. So, be patient.

For now, just understand, one Planck is the smallest energy amount a photon can differ from another photon. Physicists have measured this amount and it is one Planck. Photon energy does not increase linearly then, but in tiny Planck steps.

The suggestion is: Large photons are built of bundled god-photons, each god-photon possessing the energy of one Planck or thereabouts. Configurations, however, may be more complex—or not.

Ch. 7 – QUARKS

QUARKS -- ELECTRONS AS FURTHER BUILDING BLOCKS

If electrons and anti-electrons meet at a vector point, moving the same direction, their spins could mesh like gears. Electrons and anti-electrons could combine at a stable point. And in combinations of twos and threes or fives, electron combos could create quarks.

These quarks could have predictable vector charges. Like electrons, they would be point stable, and they could re-combine to create protons and neutrons.

Again, in the photon storm universe, God employs simplicity and beauty to design *everything*, as Einstein would have wished. By using the *simplest* possible configurations, using photons, electrons, and quarks as stable building blocks, eventually the total adds up to a rational universe.

How does the photon storm create quarks?

If we use the absolutely, positively easy-simplest configuration of the electron as a quark building block, this model seems to function quite well. (We unfortunately tried every other conceivable model first, and this took a lot of time.)

(Next page) Simply push three electrons together, two anti-electrons and one electron. The two anti-electrons unite around the electron at a shared point and rotate rather like an oil drill bit. Their spins mesh together neatly. The anti-electrons move to their positive charges; the electron moves to its negative charge. The combined unit has a vector directional charge of plus 2/3 or thereabouts. This is an up quark.

If 3 regular electrons and two positive electrons unite, the vector charge directionally forward is minus 1/3. This is a down quark. Notice how all plus and minus spins mesh nicely.

These charges and the relative angles may be further validated using vector math.

Anti quarks are as simple as quarks, and so should exist in substantial numbers with opposite vector charges if this model is correct.

THE SIMPLEST MODELS WORK BEST. THANK YOU EINSTEIN. THANK YOU GOD.

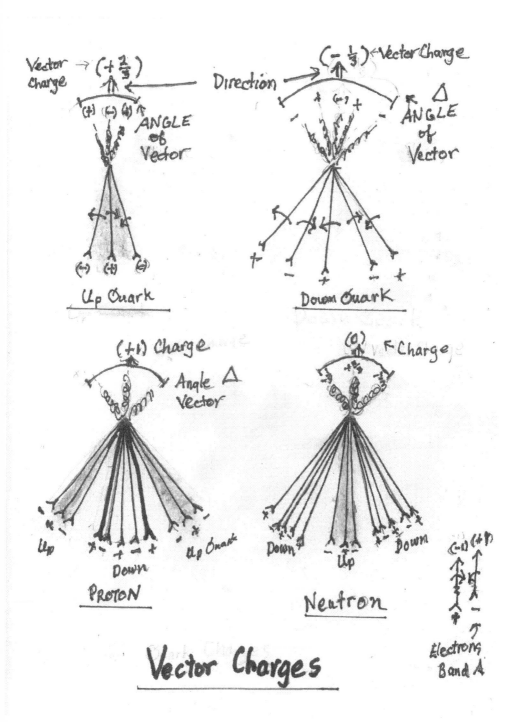

Vector Charge → (+ ⅔)

Direction →

(- ⅓) ← Vector Charge

ANGLE of Vector

Up Quark

Down Quark

(+1) Charge

Angle △ Vector

Up Quark

Down

PROTON

(0) ← Charge

Down

Up

Down

Neutron

Electrons Band A

Vector Charges

Ch. 8 –
PROTON NEUTRON CREATION AND COMBOS

HOW DO QUARKS CREATE A PROTON?
IN THE PICTURES:

Unite two up quarks with one down quark and the combined particle has a vector charge of plus 1. This particle is a proton.

Most of the proton's mass is provided by Bourne-fuzz. The proton is an extraordinarily stable and simple particle. The power of the photon storm is pushing the proton configuration together forcefully. Not much in the universe can pull it apart.

Indeed, simplicity and stability is the proton's reason for being. Nothing in the universe is apparently more simple or stable. The proton is *it*.

However, if the proton were *not* being pushed together each moment by the power of the photon storm, it would spin apart on its own. So, in a real vacuum, the proton would be unstable and dissipate. Now, we realize why we should be thankful the photon storm is around us. Thank you!

In addition, at high speeds approaching the speed of light, the stability of the proton would decline. As speed increases, the photon flow within the proton would take longer routes, and so time in that proton would theoretically slow down compared to other protons not moving as fast. The PS model gives an idea what the flow looks like in a proton model.

We will get to the math later, but Einstein has done an excellent job. The photon storm model creates a simple visualization of the process that students may appreciate.

EVEN NUMBERS

Electron quark models combining 2, 4, and 6 electrons are possible too, but the *charges* don't seem to work out correctly. These quarks would have no charge. Models with even numbers might put electrons in opposition to their opposite twisting partner, and so particle stability *could* be a problem with even numbers. Even numbered quarks *may* be taboo.

Note also: An electron may be deflected from the particle point by its charge, but a pre-electron neutrino could move passed the protective quark arms, since it has no charge. If the electron neutrino bounces off the point of a proton, its parts could instantly be focused into an electron by aligning to the same point. Thank you. I am now a stable electron.

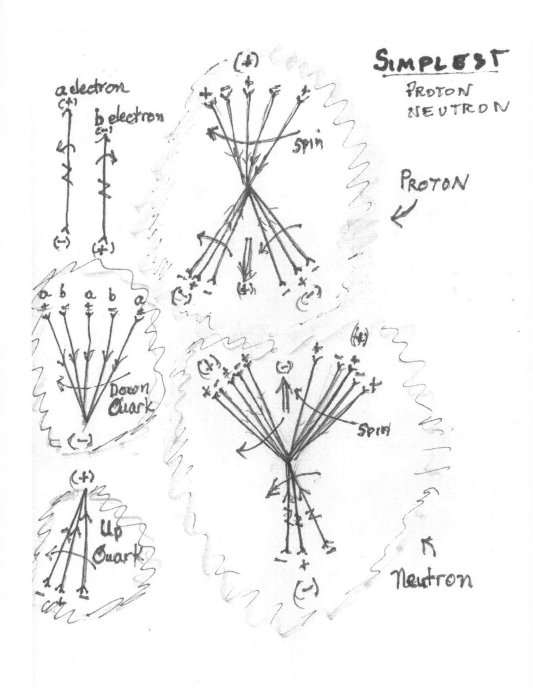

HOW DO QUARKS CREATE NEUTRONS?

If we put two down quarks with one up quark, the combined vector charge is zero. This configuration is a neutron. Again, most of its mass is supplied freely by Bourne-fuzz.

HOW AND WHY DO NEUTRONS AND PROTONS COMBINE WITH ELECTRONS?

Again, *stability* is the key. Many other particles besides protons, neutrons, and electrons are possible. These three particles are simply more *stable* and so they dominate the field. Even more important, when these particles get together, they are even *more* stable than on their own.

Please don't assume the models of the neutron and proton are absolutely accurate; but note how the spinning arms of the neutron can knock together. Not only might this spin damage the neutron, it might easily turn the neutron into a proton.

The assumption is: Two electrons could be knocked from the spinning down quark, exiting as an electron and a more damaged anti-electron neutrino with its spin reversed. The down quark would then turn into an up quark and the entire particle would turn into a proton with a positive charge.

If the neutron unites with a proton, however, the spinning arms are held apart. The proton-neutron combination is thus more stable than either particle separately. Note: When an electron attaches, the entire configuration experiences a constant flow of photon energy throughout and within.

Thus the reason for the universe is stability, not magic, not multiple dimensions, not improbability—Einstein would be pleased.

But don't rule out multiple dimensions or even magic now. Who knows what might pop up later? We don't.

Indeed, beauty is in the eye of the beholder. To us, a rational and graceful universe possesses far more beauty than an irrational universe. Some may disagree.

In the following picture, the proton and neutron—simplified versions of those pictured on the previous pages--meet at their directional points. Their quark arms are brought together by the opposing charges on the arms, leaving their middle arms momentarily wagging free.

When an electron arrives, guided by its negative directionality, it meets the positive charged quark middle arm of the proton. The trailing positive charge of the electron is attracted to the free negative middle arm of the neutron.

The final configuration allows a flow of charge completely through the deuterium atom, aided by the attached and bending electron. Not only charge, but also a flow of fuzz can circulate completely through the combination giving the proton-neutron configuration a great deal of stability. Why doesn't the free electron join at a quark point? Perhaps the charges of the quarks deter an easy addition. Or perhaps some taboo forbids even-numbered quarks.

This proton neutron configuration further acts as a building block to paste together larger atoms. As atoms are forced together, they shed some of their Bourne-fuzz as energy. The reason this fuzz must be released is because the atomic configurations become more tightly wound as each proton-neutron pair is added.

(i)

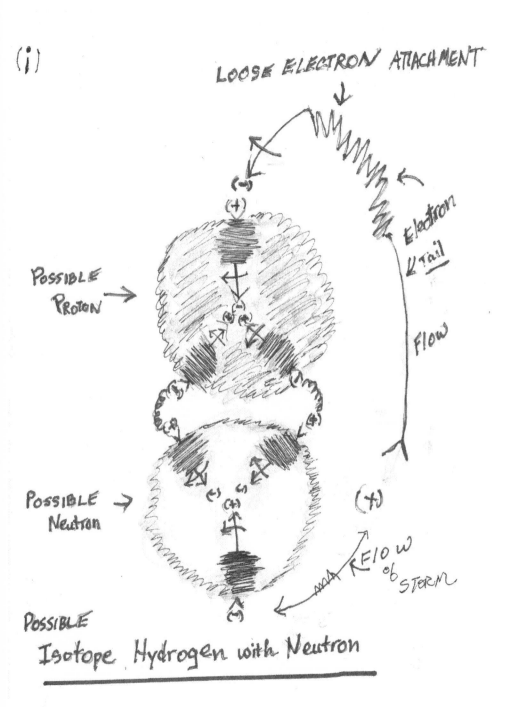

LOOSE ELECTRON ATTACHMENT

Electron
Tail

FLOW

POSSIBLE PROTON →

POSSIBLE Neutron →

FLOW of STORM

POSSIBLE
Isotope Hydrogen with Neutron

How does the photon storm create atoms?

And if the universe is rational, then we should go ahead and assume protons and neutrons assemble with electrons in *reasonable and stable* atoms and molecules.

Ch. 9

BUILDING ATOMS AND DEFINING ELECTRON ARCS

These elemental particles are *still* quantum in nature and capable of mixed personas. But in the PS universe, stable personalities dominate, and so we can *visualize* these configurations with some confidence they exist. We do not need to depend totally on math models.

Imagine you have a number of friends with split personalities. Fortunately, most of your friend's personalities only occupy them for microseconds. In addition, your friends all have *one* stable personality that stays constant for an hour or so.

If you drop in on a party of your friends with split personalities, on average you will find everyone to be in their normal stable happy self. If you look closely, you may note an unstable personality flashing for a microsecond or two in one of your friends, but the stable personality will reappear over time. In fact, if you don't look too closely, you may never know anything is amiss.

So even though unstable personalities exist, you and your friends can enjoy your gathering in normalcy and reason. The same situation is true in parties of atoms. Stability dominates over time, assuming configurations have stable states.

So, in assembling elemental particles into atoms and molecules, we can *assume* stable states exist, and if they exist we can construct *rational* models with confidence *they* exist too.

Further, if atoms assemble in simple, reasonable ways, then these configurations should predict electron orbitals. Note the electron does *not* orbit the atom. Rather, the electrons attach balloon-like to the proton-neutron configurations. The electron does not create an orbital; rather, it adapts itself to whatever space it can get, moving back and forth at near the speed of light in an arc.

In other words, the shape *inside* the atom must define the shape of the *outside* electron orbitals.

In the case of the deuterium atom the electron is free to move in a 360-degree arc around the atom.

Two electrons must share the space in helium in picture (j). These electrons can select the closest routes, while having opposite directionalities. The helium atom is pulled together tightly due to the packing of its components and the short electron orbits. Note in reality the arcs of the electrons might be quite a bit larger and the atom very, very *tiny* in comparison to the size of the electron arcs.

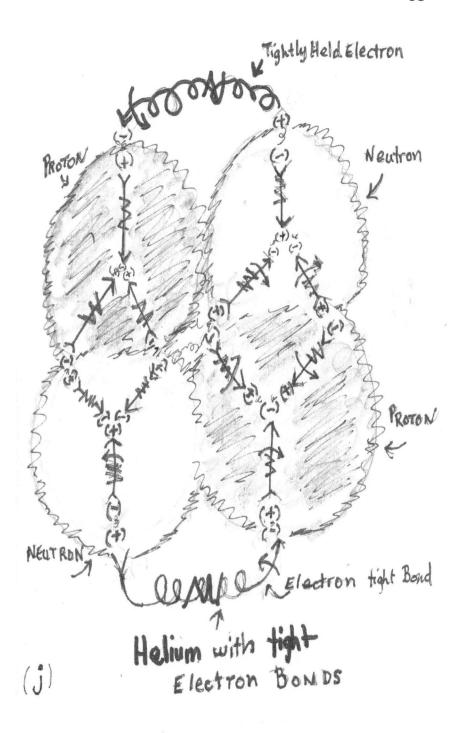

Tightly Held Electron

Neutron

PROTON

Neutron

PROTON

NEUTRON

Electron tight Bond

Helium with tight Electron BONDS

(j)

If proton-neutron pairs combine in the *simplest* possible configurations, considering their mutual charges, then the following configurations may be *somewhat* realistic and useful to predict orbits and bonding.

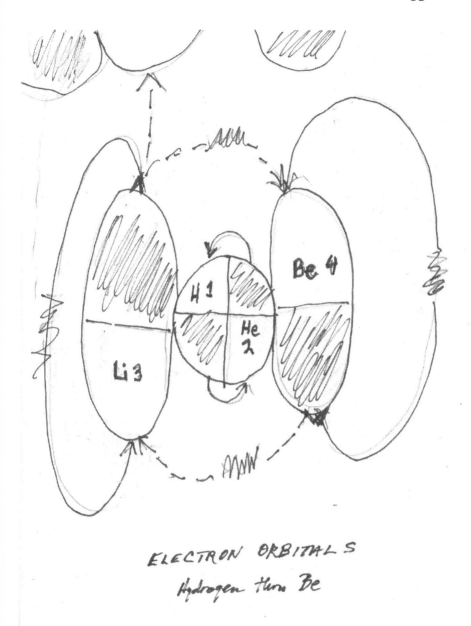

ELECTRON ORBITALS
Hydrogen thru Be

M)

**Picture M shows the possible electron arcs
from hydrogen through beryllium. Note the
arcs can be outside from pole to pole or**

alternately stretch between neutron-proton pairs. The wide arc of the lone lithium would definitely be large and capable of stretching to a separate atom, as would the beryllium orbit.

SIDE VIEW Hydrogen thru NEON

TOP VIEW NEON

Mg 12   Na 11

N)

The picture N, above, is the *side* view of the inside of a
nucleus, adding proton/neutron pairs one by one
through neon. Note the side arcs of some pairs would

invite interactions with other nearby nucleuses. However, the tenth pair, creating neon, very tightly binds the neon core. Such a neon configuration would be very reluctant to react chemically.

Sodium and magnesium pairs would be forced to attach outside the neon core, and so their electron arcs would be *much* freer to form chemical reactions.

The following picture, O, shows the inside of a nucleus as seen from the top and bottom. The proton/neutron pairs appear with the top pair partner as seen from the top, and the bottom pair partner as seen from the bottom. This view attempts to demonstrate how the different electron arcs flow through and around the nucleus. S arcs occur between numbers 1-2, and 3-4 pairs. The P arcs pair 5-7 with 8-10 pairs.

The placements of the proton-neutron pairs were not instinctive, but, after some re-positioning, these placements seemed to make the best sense. These placements also reasonably predict the shapes of the s and p electron arcs.

The teaching point is: Such model configurations give valid and predictable information to those who wish to study nuclear phenomena. They should prove helpful in the classroom.

No reason exists why further photon storm models cannot be drawn with predictable relations to heavier nucleuses. These should also be more informative than rote memorizations for students studying electron orbitals and the chemical reactions these orbitals create.

Quantum mathematical models are difficult to teach since most students and few teachers have the time to master quantum repercussions. A visual model is much simpler to comprehend for everyone. Photon storm models also generate math models that are easier for everyone.

ELECTRON ORBITALS 1-10   TOP and BOTTOM

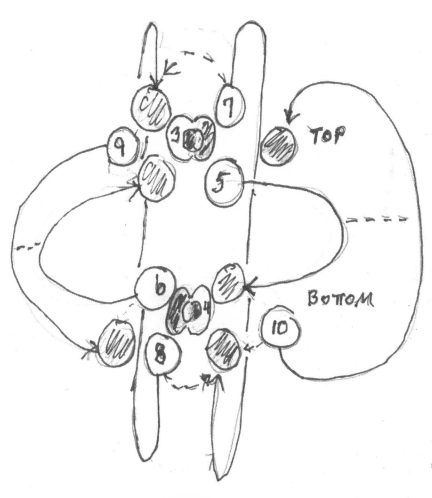

TOP

BOTTOM

NEON 10

0)

THE *GANG* OF POSSIBLE PARTICLE CONFIGURATIONS

The Gang

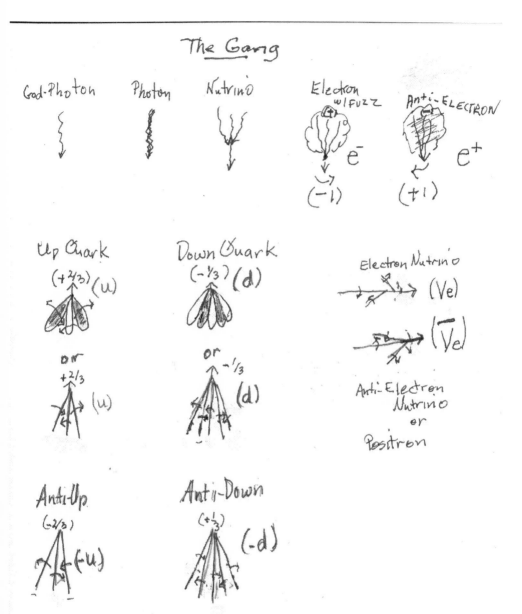

God-Photon

Photon

Nutrino

Electron w/ FUZZ

e^-

(-1)

Anti-Electron

e^+

$(+1)$

Up Quark

$(+2/3)$ (u)

or

$+2/3$ (u)

Down Quark

$(-1/3)$ (d)

or

$-1/3$ (d)

Electron Nutrino

(Ve)

$(\overline{Ve})$

Anti-Electron Nutrino or Positron

Anti-Up

$(-2/3)$

$(-u)$

Anti-Down

$(+1/3)$

$(-d)$

Ch. 10
THE EDGE OF THE UNIVERSE

In the classic universe, contemplating the edge of the universe is difficult, since most physicists feel the universe is infinite in size. The classical universe thus may *have no* edge to contemplate.

But the photon-storm universe should have an edge worth contemplating. In the PS universe we can assume: ENERGY ~ SPACE. Energy *defines* space. Energy defines space perhaps like air blowing up a balloon. We have now returned to Einstein's cosmic constant. We can assume—whether the universe is expanding or diminishing—it has an edge.

<u>How much energy is necessary to define space?</u>

If 0 energy can define space, then space can be infinite.

If greater then 0 energy is necessary to define space, then space is defined by the amount of energy in the universe.

And finally, if the cosmic constant is constant, is this amount of energy even or variable throughout the universe?

Imagine you are blowing up a balloon. It takes a little extra energy to get the balloon started. The rubber in the balloon pushes back on the air in the balloon. The question is: Does space have a similar type of elasticity? Does space push back on energy? Or is space inert, leaving everything up to energy?

Below is a view of the edge of the universe. The edge has a width. At the inner edge the intensity of the storm is 100%. Matter can exist and live inside this edge. At the outer edge of the universe, the storm intensity is zero. No matter exists at 0 storm intensity.

So we repeat, between its inner and outer rims, the edge has a width. But we don't know what this width is. The width of the edge of the universe—zero to infinity--actually depends on the elasticity of space and/or the nature of energy.

(P)

The Edge of the Universe
⬇

Inside Edge   Outside Edge

Photons
↓   ↑
                                                          (The)
                                                            Nothing
                                                              ↓

                                                          (0% Storm)

                    100% ← Edge Width → 0%

(100% Storm)

Photons
↓

ENERGY MAY ALSO DEFINE SPACE INTERNALLY

If energy confines itself, then the size of the universe would still be finite, without regard to the elasticity of space.

IS THE PS UNIVERSE SELF-CONFINED?

In the PS universe, photons do not move very well in a vacuum, in fact they probably can't move at all.

So, if we assume the edge of the PS universe is surrounded by a vacuum, then photons reaching this edge would likely stretch out blindly, and when they find nothing, they would fall back and flow together with their friends around the edge.

So the edge of the universe might be self-confined and the entire PS universe might also be defined and finite.

In other words, space would be defined by energy. We could further assume: Energy *is* space. When Einstein states space is warped by gravity, he is absolutely accurate—only the space he describes is not empty space; it is space defined by a necessary amount of energy. Thus, without energy, space may not exist.

Taking a separate angle, assuming a vacuum exists outside the PS universe, how can we measure it? If the Hulk tosses a baseball at the PS edge, the ball will dissolve at the edge and go with the flow around the perimeter. If we stick a ruler in the edge to measure infinity, the ruler, too, dissolves, unable to be solid when not assisted by the energy of the storm. A lazar measuring device also fails us since lazar beams are simply photons also and cannot move outside the storm.

After using all possible methods to measure the imagined infinity outside the edge of the universe, we must conclude infinity is immeasurable. More important, we must conclude that infinity occupies no space. Finally, we realize, a nothing that occupies absolutely no space doesn't *exist*—not in our dimension anyway. If space doesn't exist, then it isn't there at all. If it isn't there, then it doesn't exist. If infinity does not exist, then— until further notice--the PS universe occupies a finite amount of space. And in conclusion, nothing that takes up no space, isn't there.

Hopefully, that clarifies everything.

IS THE ENTIRE UNIVERSE EXPANDING OR DIMINISHING?

And since space is a function of energy, the question of whether the universe is expanding or not, depends on whether the amount of free energy in the universe is expanding or diminishing. And we must still wonder if a minimal amount of energy is required to define space, and what this amount is.

We will discuss black holes in a moment. Apparently black holes have the unique ability to expel and absorb space. The variables of the black hole phenomena may be a determining factor in the expansion or diminution of the universe.

Ch. 11
AND HOW ABOUT THOSE BLACK HOLES?

We recognize black holes possess lots of gravity, but in the PS universe, we know gravity is supplied by the photon storm; gravity is only indirectly a product of matter. The black hole matter absorbs energy from the storm, so other matter is pushed to the black hole from a higher to a lower intensity. Still, the energy of the storm is doing all the pushing.

Perhaps a black hole acts like a drain in a tank of water. The photon pressure pushes to a perceived outlet. But the exact photon configuration that enables a black hole is not clear. This concept is more easily rendered mathematically. We can introduce unknown variables to cover our lack of a concise picture.

Utilizing a visual model, the question is--do photons enter the black hole domain and become entangled forever? Or, do particles enter the domain and spin into photons that are later radiated as dark energy/god-photons?

Simply because a black hole gives no light does not mean it does not expel a great deal of energy. The quantity of this energy is unknown, but if it is in the form of God-photons dark energy we could not easily measure it.

FLOW IN = (S)(Variable) (r3/r2)

FLOW OUT = (S) (Variable2)(r2/r3)

Both flows in and flows out of a black hole depend on the power of the photon-storm (S), a variable, and the cubic area of the black hole divided by the area squared of the holes surface.

An initial question regarding flow is whether a black hole is in balance or not—does the outflow equal the inflow? Are the two processes separate?

Assuming the processes are separate, then the inflow varies by the amount of matter in the collection. The storm pushes matter like large stars toward the black hole. The difference in push within the gravity shadow is enough to pull atoms apart or allow atoms to spin apart. As individual photons, this energy cannot be seen past the horizon, going in or out.

The black hole can gain mass faster than it expels it as energy, however, it almost certainly must expel energy.

In a universe in which photons depend on each other for movement the opposite direction, if the outflow clogs up substantially, the inflow might also be affected. Though incoming photons may be trapped by the flow, outgoing photons might not be constrained, particularly God-photons with near zero diameters.

The actual limits on photon flow can only be estimated. Can the flow be squeezed until it stops flowing, or does that pinch create a new dimension?

If the S force is even throughout the universe, even in black holes, then we have a simple scenario. If the S force changes radically, or rather, if the S force can maintain stability with space at different intensities, then the scenario is less simple. However, in either case, the phenomena are more predictable than in the classic universe.

We can all think more on this.

THE END OF A BLACK HOLE

Though we are uncertain on the exact configuration of a black hole, we are rather confident about its end.

A black hole should end in two ways. It might simply evaporate, if its outflow is greater than its inflow—assuming evaporation takes place.

Or, if the black hole gets old enough and big enough, might it soak up the photon storm around it completely? If enough of the photon storm disappears outside a black hole, the black hole would no longer possess gravity. Gravity, in the PS universe, is a process of the photon storm and *not* a power of matter.

Without the storm to push it into shape, a black hole would simply spin apart. Spin is the weak force in the PS universe. This spinning apart might be orderly or it might take the form of a large bang.

And the black hole might spin apart in pieces, forming waves of semi-matter that reflect how that black hole originally formed, or it could create an evenly distributed storm of photons, a photon storm.

QUESTIONS?

Ralph Bourne
ralphbourne@sbcglobal.net

The attempt of this presentation is to give as accurate a *visual* model of the phenomena of Einstein's mass, gravity, and matter as possible. The math presentation follows:

Ch. 12 –
EASY MATH AND PHOTON ENERGY

SOME EASY MATH.

Phenomenon in a universe possessing a cosmic constant or photon storm is generated by an entirely different origin than in the non-photon-storm classic universe. The math, however, is essentially the same. The formulas that evolve in the PS universe fortunately distill into formulas already created by Newton, Einstein, and friends. These formulas already give accurate results.

Moreover, whether we view god-photons as real particles or as quantum probability clouds, the mathematical results are similar—a universe of shared energy.

The primary difference is, the photon storm model can generate visual constructions of atomic activity that the classic model and the quantum model do not. It is interesting how similar formulas derive from different models.

We should start at the beginning. In a photon storm model we can begin actually with the smallest particles and graduate to larger and larger units.

PHOTON ENERGY

We know the energy of a photon has a relationship to Plank's constant. (If the reader is confused about the history of Planck's constant, a description is supplied in the back of this book, or an explanation is available online.) But we know photon energy does not grow smoothly; when measured, photon energy grows in tiny intervals.

Each tiny interval has been carefully measured, and this tiny amount is called Planck's constant. Planck's constant has a relationship to what scientists measure as a photon's wave frequency. So the classic formula for photon energy is, Energy = (Frequency) times (Planck's constant). Energy is equal to frequency times Planck's constant.

In the classic universe, photon waves are assumed to possess frequencies, waves per second, and wavelengths, the actual lengths of a photon waves from peak to peak. Photons may possess these attributes also in the PS model.

However, in the PS model, we suspect that photons are actually bundles of individual god-photons spun together in twisting threads. These threads resemble spiral waves, possessing similar measureable attributes. A photon has some stability through space, but if is absorbed, it can change back into its individual parts of pure energy.

(AMPLITUDE: In the classic universe, photons are often also ascribed amplitude as an attribute. But it is commonly agreed that photons do *not* possess amplitude. The measure of photon amplitude is actually the measurement of photon intensity, or the *number* of separate photons. So an individual photon, either in the classic or PS model, does *not* possess amplitude. Thus when we notice light growing brighter or dimming, we are noticing the *number* of photons to hit our eyes, not a changing attribute of individual photons.)

So the relationship to Planck's constant must also exist in tiny god-photons. This relationship may be a one-on-one relationship with each photon having one plank of energy, or the relationship may be more complicated—more than a single god-photon may be aligned together to equal the energy of one Planck. A configuration constant (cc1) is necessary in case the relationship is complicated. If the configuration is not complicated, then the constant would simply equal 1.

If a single god-photon energy is = G, then G is equal to (Planck's constant) times this (alignment constant). This is the energy of a *single* god-photon. G = (Planck's constant) (Alignment constant).

PHOTON ENERGY

The energies of photon *bundles*, photons in the light or radio spectrums, are simply equal to the (*number* of g-photons) times (*G*). E = (Number) (G) = (#) (G).

In the classic model: Energy = (Frequency) (Planck's Constant). The energy of a photon is the frequency of the photon time Planck's constant. The assumption in the PS model is, when scientists measure frequency, they are *actually* measuring number (#).

So photon energy in the PS model is simple multiplying G times the number of photons, or E= (G) (#).

Remember, photons also have spin, + and -. Spin is utilized by all photons so they can easily move past each other. Remember also, that an electron has spin in common with photons.

ELECTRON ENERGY

The formula (G) (#) (cc2) works also when assessing the mass/energy of an electron. The concept is this: An electron is simply a stable photon configuration. The electron's configuration attracts fuss-mass and keeps the electron from moving like a normal photon at light speed. We need a second **configuration constant** for an electron to denote the (#) amount of fuzz that attaches related to mass. Call this constant (cc2).

To restate, we know a relationship must exist between the absolute number (#) of photons aligned with an electron and its mass, but this relationship may or may not be direct. The alignment (fuzz) constant is necessary to denote the ratio of this relationship. If the relationship is direct, then the value of the alignment constant would be (1).

Furthermore, the energy and mass of any electron changes. By absorbing additional photons, the energy of an electron increases (the velocity of the electron would also increase), thus the mass of an electron would vary with its energy-velocity.

If course, an electron cannot move faster than light. Its configuration interferes with the photon storm (and that is why it has mass and photons don't).

Like a photon, an electron also has clockwise or counterclockwise spin. We have arbitrarily given a clockwise spin to an ordinary electron, which creates a negative (-1) directionality. The counter-clockwise anti-electron has a (+1) directionality. Spin, however, may actually be reversed.

If we divide the energy mass of an electron by Planck's constant, we have *some* idea of the number of photons in an electron. If an electron's rest energy/mass is 511,00 Evs, and the energy of Planck's constant is 4.1357 X (10)-15 Evs, then we get something like 123558 x (10)15 photons in one electron. 123558.28 *x* (10)15 is quite a few photons. We still need to assume an alignment constant (a2) which may not be one (1), but the number would probably be large in any case. $123558 \, x \, 10^{15}$ or $1.23 \, x \, 10^{10}$ or thereabouts god-photons in every electron. Of course this number would increase or decrease as the electron receives or emits energy and simultaneously changes velocity.

Mass is created directly from energy particles. So when mass reconfigures back to high-energy photons, the energy must be the same as the total number of these particles divided by the original configuration constant, which may or may not be one.

So E= MC2 = (#) (PLANCK'S CONSTANT) /(cc2)= (#) (G).

Ch. 13 --

CHARGE AND ELECTROMAGNATISM

And most important, a spinning electron turns the photons moving through it into powerful funnels in both directions. Opposite twisting funnels interlink and pull on each other. The clockwise spinning electron has arbitrarily been assigned a (-) charge, and the counter-clockwise anti-electron has a (+) charge in line with the direction of movement.

Charge is carried by spinning tornadoes of god-photons pulling against or pushing away from other tornadoes. In the PS universe we can draw simple pictures of this force. If the charge of a single electron at rest is C, then the charge of 10 electrons would be 10 C or (# C). The force between two electron groups pointing at each other would be (Alignment constant)(#C1) (#C2)/d2. With the introduction of a charge constant, this turns into Coulombs formula: F = (Coulomb's constant} (C1)(C2)/r2. Force between the charges equals the charges multiplied together times (Couloub's constant) divided by the distance between the charges squared.

ATOMS ARE CREATED FROM ELECTRONS THAT ACHIEVE POINT STABILITY.

THE PHOTON STORM PUSHES ELECTRONS TOGETHER AT A POINT. THE FORCE OF THE PHOTON STORM IS ALSO KNOWN AS THE STRONG FORCE.

SPIN IS THE WEAK FORCE THAT CAUSES ATOMS TO TEAR APART.

Quarks are made up of electrons and anti-electrons side by side because their opposing spins mesh together without conflict. However, if an electron meets an anti-electron, head to head, they might annihilate each other. If an electron runs head to head with another electron, they would rudely push each other aside. But, if an electron is shoved *forcefully* together side by side with another electron, the spins would contradict violently. This side-by-side annihilation is what happens when a neutron turns into a proton.

Ch. 14 --

THE CHANGE OF A NEUTRON INTO A PROTON.
AN EXCELLENT EXAMPLE OF HOW THE WEAK FORCE
CAUSES ATOMIC DECOMPOSITION

THE NEUTRON TO PROTON CHANGE HAPPENS MOST
OFTEN IF A NEUTRON IS ENERGIZED. THE TWO DOWN-
QUARK ARMS OF THE NEUTRON CLASH TOGETHER
FORCEFULLY. THE ELECTRONS ON THE ENDS OF THE
QUARK ARMS ARE SQUASHED TOGETHER WITH THEIR
SPINS CONFLICTING. NORMALLY FREE ELECTRONS CAN
AVOID CONTACT WITH EACH OTHER, BUT THE ELECTONS
ON THE QUARK ARMS ARE ATTACHED RATHER TIGHTLY.

SO WHEN THE TWO ELECTRONS CLASH, THEIR SPINS
DIG DEEPLY INTO EACH OTHER TEARING EACH OTHER
APART. THE CONCUSSION IS ENOUGH TO KNOCK BOTH
ELECTRONS FROM THEIR ATTACHMENTS WITH THE
DOWN QUARK. ONE ELECTRON LIMPS OFF CRYING.   THE
OTHER ELECTRON IS SO INJURED IT LOSES ITS
ATTENDING FLUFF ENERGY AND ITS SPIN IS TOTALLY
REVERSED. IT IS NOW AN ANTI-NUTRINO.

A SECOND ELECTRON IS ALSO MOMENTARILLY
SEPARATED FROM THE QUARK. AS A FOURSOME IT DOES
NOT FIT—FOURSOMES ARE APARENTLY TABOO--BUT AS
THE WOUNDED ARMS SPIN AROUND, IT IS PUSHED TO A
FIFTH END POSITION ON THE ARM NEXT TO IT. THE
ADDITION OF THIS ELECTRON CHANGES ONE 4-
COLLECTION BACK INTO A 5 DOWN QUARK AND THE
OTHER SET INTO A 3-ELECTRON UP QUARK.

THE REMAINING PARTICLE IS NOW A PROTON.

(See pic, next page)

SPIN, *THE WEAK FORCE*, CAUSES THE DISASTER.

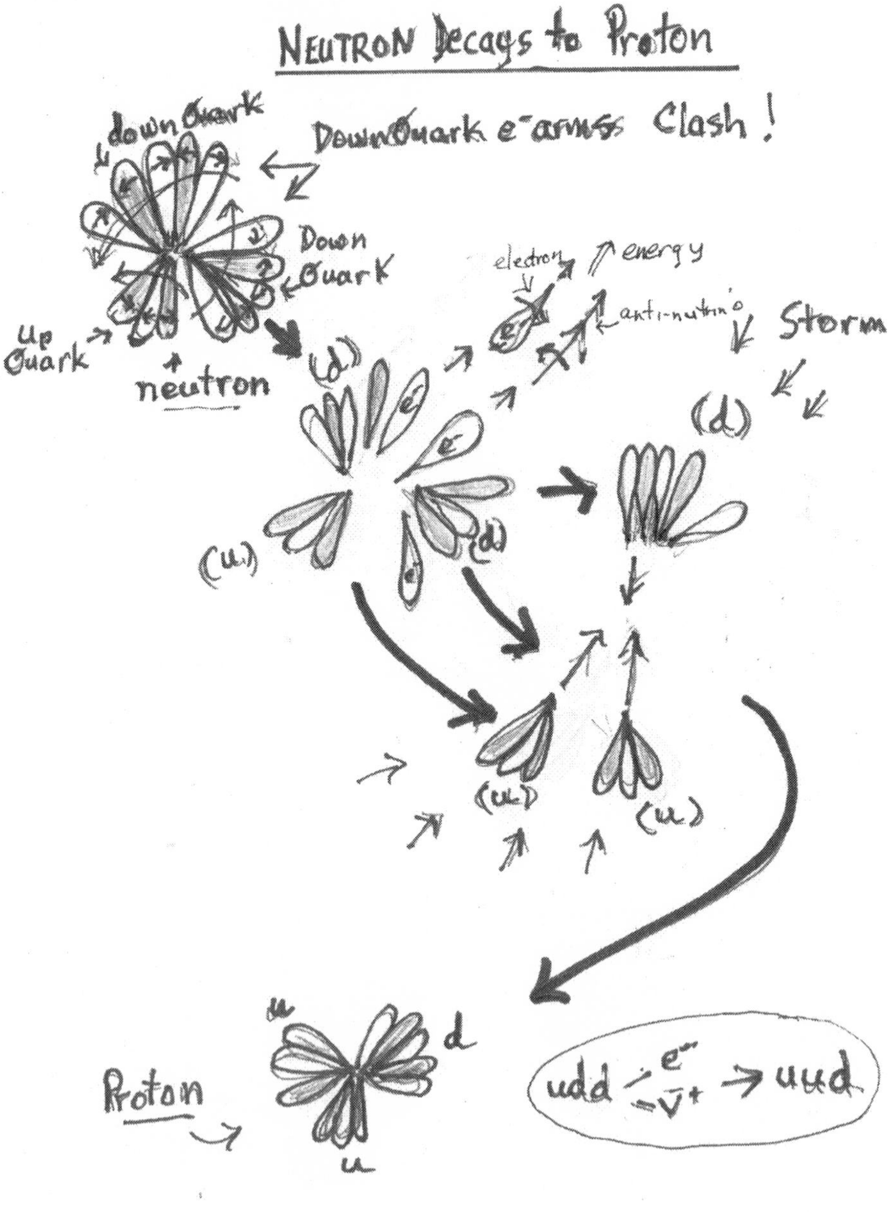

NEUTRON Decays to Proton

DownQuark e⁻arms Clash!

down Quark

Down Quark

Up Quark

neutron

(d)

electron / energy

anti-nutino° Storm

(d)

(e⁻)

(e⁻)

(u)

(d)

(u)

(u)

Proton

u

d

u

udd → e⁻ / ν⁺ → uud

Ch. 15
SPIN, *THE WEAK FORCE*

NEUTRON DECAY DEMONSTRATES THE INCREASING DIFFICULTY OF CONFIGURING STABLE ATOMIC PARTICLES AS THE SIZE AND NUMBER OF QUARKS INCREASES. AS THE CONFIGURATIONS GROW MORE COMPLEX, THE MORE VARIETIES OF SPIN, THE MORE CHANCE OF OCCASIONAL DISASTERS WHEN THESE SPINS INTERACT.

THE SPIN DECAY FACTOR INCREASES SIMILARLY IN VERY LARGE ATOMIC NUCLEUSES, COMPOSED OF MANY PROTONS AND NEUTRONS. THE MORE VARIETIES OF CONFLICTING SPINS, THE BIGGER THE CROWD, THE MORE PROBABILITIES OF A CRASH.

THE PROBLEM IS THE SAME AS ON A FREEWAY. IN BOTH CASES, SPEED AND SPIN ARE THE PROBLEMS AND THESE INCREASE WITH PROXIMITY.

IN CONCLUSION: PARTICLES DO NOT DECAY BECAUSE OF PHANTOMS OR GHOSTS. THE REASONS ARE QUITE MECHANICAL INSTABILITY INCREASES WITH INCREASING OPPORTUNITY FOR COLLISIONS AS NUMBERS INCREASE IN A SMALL NUCLEUS. SPIN IS THE WEAK FORCE THAT TEARS NUCLEUSES APART.

RECALL, HOWEVER, SPIN, MASS, AND GRAVITY ARE ALL RELATED PRODUCTS OF THE PHOTON STORM. IN THE PS UNIVERSE, THE WEAK FORCE IS NOT AN ORPHAN.

CH. 16

CHANGING A PROTON INTO A NEUTRON

THE PROTON IS MORE STABLE THAN A NEUTRON, IN PART BECAUSE ITS ARMS ARE SIMPLY SMALLER AND LESS LIABLE TO RUN INTO EACH OTHER. SO DECOMPOSITION OF A FREE PROTON IS UNLIKELY. HOWEVER, A PROTON CAN AFFECTIVELY CHANGE TO A NEUTRON IN SEVERAL WAYS IF IT IS CROWDED IN TRAFFIC AND THE ENERGY IS HIGH.

ONE POSSIBLE WAY IS IF THE QUARK POINT IS STRUCK BY AN ANTI-NUTRINO. THE ANTI-NEUTRINO IMMEDIATELY TURNS INTO AN ANTI-ELECTRON WHEN ITS POINTS ALIGN. HOWEVER, THIS NEWLY CREATED ANTI-ELECTRON FORMS ON A POINT EXACTLY OPPOSITE AN ELECTRON. THE TWO PARTICLES WOULD GRIND INTO EACH OTHER AND ANNIHILATE IN A BLAST OF ENERGY THAT WOULD UPSET THE ENTIRE QUARK. HOWEVER, IF AN ANTI-UP QUARK WAS AVAILABLE TO FILL IN THE HOLE, A NEW DOWN QUARK WOULD FORM, AND THE NEW PARTICLE WOULD BE A NEUTRON AFTER THE DUST SETTLES. THIS DECAY PROBABLY DOES NOT OCCURE IN NATURE.

A SECOND, MORE OBVIOUS WAY A PROTON COULD DECAY INTO A NEUTRON WOULD OCCUR IF THE PROTON WERE CROWDED AND ENERGIZED SO THAT ITS UP QUARK ARMS SLAMMED INTO EACH OTHER LIKE THOSE IN A NEUTRON. THE DECAY, HOWEVER, WOULD BE THE OPPOSITE OF NEUTRON DECAY. ONE WING WOULD LIMP OFF AS AN ANTI-ELECTRON AND IMMEDIATELY ANNIHILATE. A SECOND WING WOULD BE DAZED BUT ABLE TO RECOVER.

FINALLY, THE SURVIVING WINGS WOULD REASSEMBLE, WITH THE AID OF AN ANTI-UPQUARK, INTO A NEUTRON.

THE ARMS WOULD BE PUSHED BACK TOGETHER BY THE POWERFUL FORCE OF THE PHOTON STORM. THE ANTI-UPQUARK MIGHT BE SUPPLIED BY THE DECAY OF AN ASSOCIATED NEUTRON. THIS NEUTRON DECAY MIGHT BE AN INITIAL STIMULANT FOR THE PROTON DECAY IN A CROWDED NUCLEUS.

SUCH A DECAY MOST LIKELY OCCURES IN NATURE.

ELECTRON CAPTURE

IN A THIRD POSSIBILITY, A PROTON COULD TURN INTO A NEUTRON IF AN ELECTRON NEUTRINO, HAVING NO CHARGE, COULD SLIP PAST THE CHARGED ARMS AND REACH THE POINT. THE ELECTRON NUTRINO MIGHT REALIGN INTO AN ELECTRON, AND CREATE A DIVERSION THAT ALLOWS THE CIRCULATING ATOMIC ELECTRON TO ATTACH ALSO. THE PROTON IS NOW A NEUTRON.

SUCH A SITUATION SHOULD BE VERY RARE IN ORDINARY NATURE. A NEUTRINO-ELECTRON THAT

STRIKES A PROTON USUALLY WOULD NOT AFFIX, BUT WOULD BOUNCE OFF PERHAPS AS AN ELECTRON. ONLY IN A VERY INTENSE PHOTON STORM WOULD THE CHANGE POSSIBLY TAKE PLACE, FOR EXAMPLE, IN A NEUTRON STAR.

THE IMPORTANT FACT IS: PHYSICAL MODELS USING THE PS MODEL CAN GENERATE THE SAME RESULTS AS THOSE GENERATED BY QUANTUM MECHANICS—AND THE CONCLUSIONS ARE MUCH SIMPLER TO UNDERSTAND.

The photon storm model allows the creation of numerous other configurations—variant neutrinos, pions, muons, plus and minus. These configurations are usually not stable to the degree they can form long-term matter. However, they may have roles during transitional periods of ordinary matter.

However, in the PS model, stability determines almost everything of weight, mass, and gravity.

Ch. 17
MATH OF MASS AND GRAVITY

Mass and Gravity Math

What is mass in the PS universe? Mass equals the power of the storm pushing on a particle to create inertia. Remember the atom named Bill. The power of the storm we will call (S) or STORM. If we multiply the power of the storm (S) by the cubic-area (A3) of a particle, and multiply this by the reflectivity (Rf) of the particle, we end up with a number, which is the absolute mass/inertia of that particle. Inertia is the force the storm exerts on a particle to keep it in the same place. Remember Bill again. A reflectivity variable (Rf) is necessary because a particle only reflects a tiny amount of the energy of the storm.

In common large bodies, like human beings, and planets, this tiny, tiny variable is constant. However in extreme cases, like large stars and black holes, the (Rf) variable may change. These bodies may reflect or absorb a larger amount of the photon storm energy than regular bodies.

So to repeat, the mass of a normal body equals the force of the storm (S) times the cubic-area of the particle (A3) times the normal reflective variable (Rf). **MASS = (S) (A3) (Rf)**

GRAVITY

Gravity in the PS model occurs because bodies that possess mass possess that mass because they reflect the intensity of the photon storm. Thus all particles with mass create gravity shadows between themselves and other particles. Remember the picture of the space couple in the rain shower?

GRAVITY.

TWO OBJECTS SHIELD EACH OTHER AND ARE PUSHED TOGETHER BY PHOTON STORM

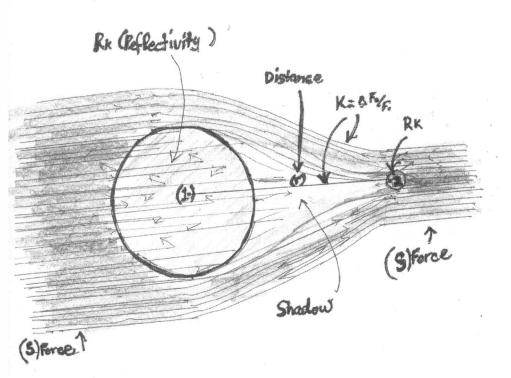

The force pushing these objects together, the acceleration would be supplied by the photon storm.

In a simplistic picture assume the two planets are exactly next to each other in space. Then assume each planet actually is the size of a point. The storm would be pushing on these points from both sides with the power of (S) (Rf) times the different masses (S) (Rf) (a1) times (S) (Rf) (a2). This would be the total force each object is placing on the other. When the objects are at distance the force would diminish by (r2) the distance between the objects squared.

Thus the formula in the PS model for the GRAVITY FORCE is: GRAVITY = [(S)(Rf) x (S) (Rf) (a1) x (S) (Rf) (a2) / r2. But in the PS model mass is (S) (Rf) (a), or the (S) force times Reflectivity (Rf) times cubic area (a). So the formula condenses to (S) (Rf) (Mass1) (Mass2) / r2.

If this formula is familiar, it is exactly the same as the one Newton derived many years ago for large objects in the classic universe.
FORCE OF GRAVITY = (G) (M1)(M2)/r2. We merely substitute the variable (S) (Rf) for the constant (G).

Newton assumed particles of mass *pulled* on each other. However the pulling action is the same as when particles of mass are *pushed* to each other by the photon storm. The formulas are the same even though the models for the formulas are quite different.

We now know the variable (S) (Rf) = Gravity constant. Later, with this knowledge, we should be able to compute the relative values of (S) and (Rf).

Ch. 18
GRAVITY ANOMALIES of LARGE OBJECTS

LARGE OBJECTS WITH VARYING (Rf) VALUES

Einstein recognized that very large objects bent photon energy around them in ways Newton's formula did not predict. Einstein assumed that large objects warped space-time. Such squeezing of space causes an anomaly in gravity that bends the routes of photons.

Einstein's concept is absolutely correct in the PS model also, but in the PS model the stretched space-time is not empty space, it is filled shoulder to shoulder with intense god-photon energy.

Einstein's model envisioned space to be like a mattress. When squeezed it created an emptier space in which matter fell. This gravity concept works in the PS model as well.

In the PS model, the reflectivity (Rf) variable changes in very dense objects, and may even vary significantly within even larger objects due to configuration changes. So Newton's formulas must be modified in such situations. Thus gravity anomalies in the PS model are similar to those in the classical model, the mathematical results are similar even though the model is quite different.

To repeat: Whether space-time is a space vacuum warped by mass to create gravity, or whether space-time is an energy field that creates matter and in the process warps to create gravity, the results are exactly the same.

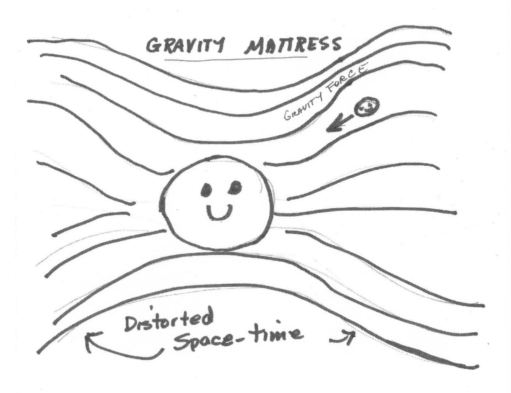

TIME -- Ch. 19
WHY TIME CHANGES NEAR
LIGHT SPEED

HOW ABOUT TIME?

Einstein's considerations for time are the same in the PS model as in the classic model. As a rocket accelerates, time slows, since the flux movements in a stretched atom take longer routes. However, these time differences only become noticeable at velocities approaching the speed of light, or in extremely intense gravity fields. These differences are more easily visualized in a photon storm model.

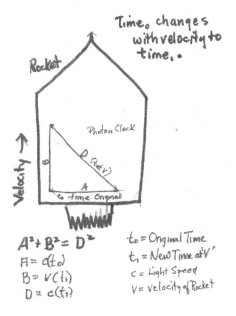

$$A^2 + B^2 = D^2$$
$$A = c(t_o)$$
$$B = v(t_1)$$
$$D = c(t_1)$$

t_o = Original Time
t_1 = New Time at V'
c = Light Speed
v = velocity of Rocket

HAPPY MATH AGAIN

For those interested in the math, here is some.

To compare the original time, To, to the time in the rocket at velocity, T1, we need to compare the length of the original clock, line A in the drawing, to the new clock length, line D in the drawing. Since the lines are connected by a right angle triangle. A2 + B2 = D2. The speed of light is (c); velocity is (v). Velocity times time = Distance.

$$A2 + B2 = D2$$
$$(cTo)2 + (vT1)2 = (cT1)2$$
$$c2To2 + v2T12 = c2t12$$
$$(To)2 = c2(T1)2 - v2(T1)2/c2$$

$$(To)2 = (T1)2 - \left(\frac{v2}{c2}\right)(T1)2$$

$$(To)2 = (T1)2\left(1 - \frac{v2}{c2}\right)$$

$$(To)2/\left(1 - \frac{v2}{c2}\right) = ((T1)2$$

$$\frac{(To)}{\sqrt{1 - \dfrac{v2}{c2}}} = (T1)$$

As v approaches c, the speed of light, T1, time in the rocket, approaches infinity.

Height varies depending on whether an object is being accelerated. A rocket accelerating will be wider and shorter like a compressed spring, but once the acceleration ends, the object will spring back in the opposite direction becoming long and thin, as if it were accelerating into a black hole. Height Original is to Height at velocity. Hto=Original Htv=at velocity.

Htv = Hto(c+v)2/(c-v)2 (K)

$$Htv = Hto\,\frac{(c+v)2}{(c-v)2}\,(K)$$

Since Mass=(S)(Rf)(A3), as (Rf) changes due to velocity, (A3) may stay the same.

Assuming the particles in the rocket will maintain their same area3, then Length1 times Length 2 times Height stays constant. The ratio of L1/L2 would also stay the same. (L1v)(L2v)(Htv)=A3

$$(L1v)(L2v) = \frac{A3}{Htv} = A3/(Hto)(c+v)2/(c-v)2$$

So, knowing original A3 and the L1/L2 ratio, the new lengths at velocity can be calculated, and they would become very small as (c-v) approaches zero and Height approaches infinity.

So the picture of an object approaching the speed of light or being sucked into a black hole is similar. The object's length would approach infinity as its widths approached zero. It would begin to resemble a photon.

Reality may have something to do with the final situation, however, which we will discuss next.

Ch. 20
WHY MASS INCREASES WITH VELOCITY NEAR THE SPEED OF LIGHT

THE RELATIONSHIP BETWEEN MASS AND VELOCITY

Mass also increases with velocity in the PS model as it does according to Einstein's calculations in the classical model.

The final equation (quite lovely) is:

$$\frac{(mo)}{\sqrt{1 - \frac{v2}{c2}}} = (m)$$

As v approaches c, the increased mass, (m) grows to infinity. (mo) represents the original mass. This is Einstein's equation, and both his equations on time and mass have been validated in the classic model.

In the PS model, the mass change should be more understandable to the reader. If fact, the mass change is a real visual change in the PS universe because the actual *number* of photons making up a particle physically increases as the particle goes faster and faster, until, near the speed of light the mass is simply too massive to accelerate. In fact, such a particle would, on reaching the speed of light, change back to its original photon energy, all god-photons again.

We can make some substitutions:

$$\frac{(mo)}{\sqrt{1 - \frac{v2}{c2}}} = (m)$$

$(S)(Rf2)((a3) = (S)(Rfo)(a3)/\sqrt{1 - v2/c2}$

Recall, in the PS model (mo) = (S)(Rfo)(a3)

ORIGINAL MASS equals (THE STORM FORCE) times (THE ORIGINAL REFLECTIVITY) times (CUBIC AREA).

NEW MASS AT VELOCITY equals (THE STORM FORCE) times (THE CHANGED REFLECTIVITY) times (CUBIC AREA).

The difference between the original mass and the new mass at velocity is dependent on the change in the reflectivity of the particle as it approaches the speed of light.

Note: We are *not* certain whether (a3) cubic area stays constant. This depends on the elasticity of space. But assuming the cubic-area (a3) stays constant makes the computations much easier. For now:

If: $new(m) = (S)(Rf2)((a3) = (S)(Rfo)(a3)/\sqrt{1 - v2/c2}$

Then, noting (S) (Rf) = #, the storm force times Refractivity (Rf) = #, with (#) number reflective of the actual number of photons in a given cubic-area, so: Assuming (a3) stays reasonably constant, we can factor out (a3) and (S) and change the equation to:

$$(Rf2)(\#new) = (Rfo)(\#original)/\sqrt{1 - v2/c2}$$

Finally, in the above formula we are working with actual numbers of physical god-photons. As the velocity of a particle increases, its Reflectivity (Rf) value increases, particularly as it approaches the speed of light.

We can reduce this again to:

$$(Rf2)/(Rfo) = (\#original)/(\#new)\sqrt{1 - v2/c2}$$

The new reflectivity compares to the original reflectivity in the same way the original number of photons compares to the new number of photons, divided by $\sqrt{1 - \dfrac{v2}{c2}}$.

So, in the PS model, we can deal with <u>actual numbers</u> of photon energy. And we can discern a relationship between v2/c2 and the value of (Rf).

As the velocity of a particle approaches the speed of light, the Reflectivity (Rf) must approach 100%. Mass approaches infinity, but in reality, the results would be defined by limits of (S), the photon storm, and the degrees to which configurations maintain stability at high velocities.

Hopefully the reader is not put off by the math and we can attempt a second, easier explanation. In the PS model, increasing kinetic energy is an *actual* increasing physical change of number. To restate: In the PS universe, god-photons actually react with a particle as its speed increases in *real* ways. The *numbers* of photons affecting mass actually increase. These photons can be *counted*.

Imagine a storm plow going through snow. As it moves faster and faster it will push more snow ahead of it. This snow makes the mass of the plow increase as the velocity of the snow plow increases.

In a similar fashion, if an electron is hit by a photon, its energy will increase as well as its speed. It was just energized *forward* by a powerful photon. The electron now moves faster forward, closer to the speed of light, but also possesses a larger *number* of photons—it is a more *massive* electron. The numbers are *real*, not invisible. But they coincide exactly with Einstein's calculations.

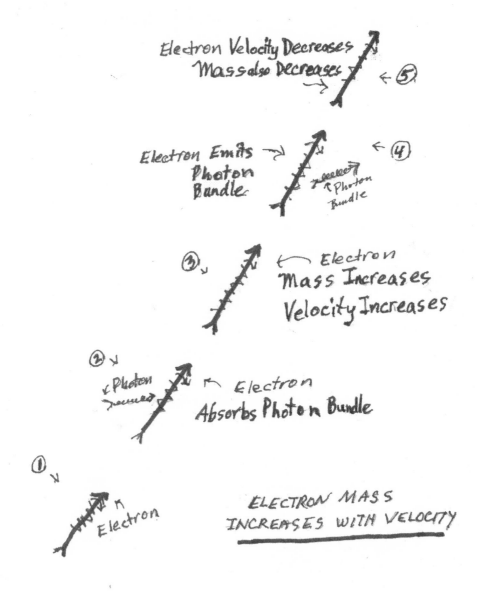

Electron Velocity Decreases
Mass also Decreases → ← ⑤

Electron Emits → ← ④
Photon
Bundle ← Photon
Bundle

③ ← Electron
Mass Increases
Velocity Increases

② ← Photon
Photon → ← Electron
Absorbs Photon Bundle

① Electron

ELECTRON MASS
INCREASES WITH VELOCITY

We have discussed that a connected
electron can only accept photons in

discrete shells, like a crystal. If an electron loses a shell, not only will its mass/energy decrease, but we should assume its speed decreases also. The shell will be given off as a photon.

An unconnected electron is more flexible in its ability to accept and emit photon energy. A free electron can accept and emit energy at many levels, and so we have electromagnetism and smart phones.

If the velocity of an electron increases, the energy of that electron increases. This energy is *real* and can be *counted* as increased photon numbers.

The closer an electron's speed approaches the speed of light; the more that electron begins to resemble a pure photon. In other words, at the speed of light, all *point configurations* would decay and the result would simply be photon energy. A famous equation comes to mind.

$E = MC2$

Thank you Mr. Einstein.

We should note, however, both (E) energy, and (M) mass are variables, *not* constants. As E increases, M increases also.

If we assume (C), the speed of light, *is* a constant in the PS model, then C2 = E/M. So energy and mass have a constant relationship to each other. They are absolutely the same thing. Only, their different *configurations* make them *appear* different at times.

So in this model, the entire universe is composed of tiny god-photons, themselves perhaps composed of even tinier energy-units—not discussed here—in different configurations.

We could argue all people are the same. Only small, petty things make us think we are at all different. Small things are only important if we allow them to divide us. Are people actually large semi-photons? That is mathematically possible.

In the PS model:

(G# Total Photons) =
[(S) (Rf)(a3)] [(c2)]

Or:

E=MC2

The two formulas are the same.

QUESTIONS APPRECIATED:
ralphbourne@sbcglobal.net

"May the storm be with you."

Thank you.

Ch. 21
"Spooky actions at a distance."
Particle Entanglements--Why Einstein may still be correct.

Einstein described particle entanglements predicted by quantum mechanics to be, "Spooky actions at a distance." Einstein did not question the general validity of quantum mechanics; but rather, he felt the described "spooky" interactions were incompletely understood.

How could the actions of one particle change the actions of a distant entangled particle faster then the speed of light as predicted by quantum theories?? Some other phenomena needed to occur that the math did not consider. What was this phenomenon? The answer must exist somewhere, Einstein felt, but it could not be seen.

As mentioned earlier, Einstein was a model person, not a math person—he utilized math as a necessary tool and not an art form. Einstein desired a complete picture; he was frustrated when the mathematics did not provide one.

Recent experiments have apparently validated the quantum world regarding quantum entanglements. However these ongoing experiments have been conducted without considering the photon storm as the required explanatory phenomenon.

If we do consider the photon storm's implications, then both Bohr and Einstein are still correct, the theory is unified, and no "spooky" actions occur.

The quantum world actually began with the observance of the double slit quandary. The photon storm model has an alternative explanation for this experiment.

Here it is:

THE DOUBLE SLIT EXPERIMENT IN THE PS UNIVERSE

What is the double slit experiment? In 1804 a physician, Thomas Young, noted light waves formed a broad even scatter when sent through a single slit; but when the same light was sent through two closely aligned slits, the resulting light patterns interfered and formed lines of light and dark.

The conclusion was, since light interfered with itself in wave patterns, light must be a wave! Great!

Later however, when electrons or photons were sent singly through double slits, they interfered in the same way as if double waves were moving through the slits simultaneously. How could single electrons or photons create an interference pattern when only one wave/particle at a time was moving through a slit?? Can a wave know in advance that it will find interference in the future??

Bohr argued for a universe in which a particle is undefined until observed; and thus this particle is free to take all possible routes to its destination--the undefined particle can subsequently interfere with itself, while defying time. Quantum mechanics originated due to the double-slit quandary.

Einstein argued for some rational, unknown explanation, but the majority of physicists, like Bohr, found adopting uncertainty principals—and asking no further questions—could acceptably solve the problem.

Physics students were required to ignore what could not be visualized and simply "do the math." And so the situation in the standard universe remains.

Due to the quantum-wave duality, physics students are advised no models can be constructed of atomic particles—something that is both a wave and a particle cannot be visualized--only abstract mathematics can explain the standard universe.

In the PS universe, however, the double slit experiment is easily explained, since the mechanism involved is easily visualized.

In the PS universe, mass, gravity, and the movements of atomic particles are all determined by interactions with the energy of the Photon Storm. In particular, the movements of photons, electrons, and molecules are determined by their spins interacting with the storm.

Therefore, in the PS universe, the movements of photons and electrons passing through slits have little to do with the photons or electrons— *the results are pre-determined by the photon-storm and the slits!!*

(See following picture)

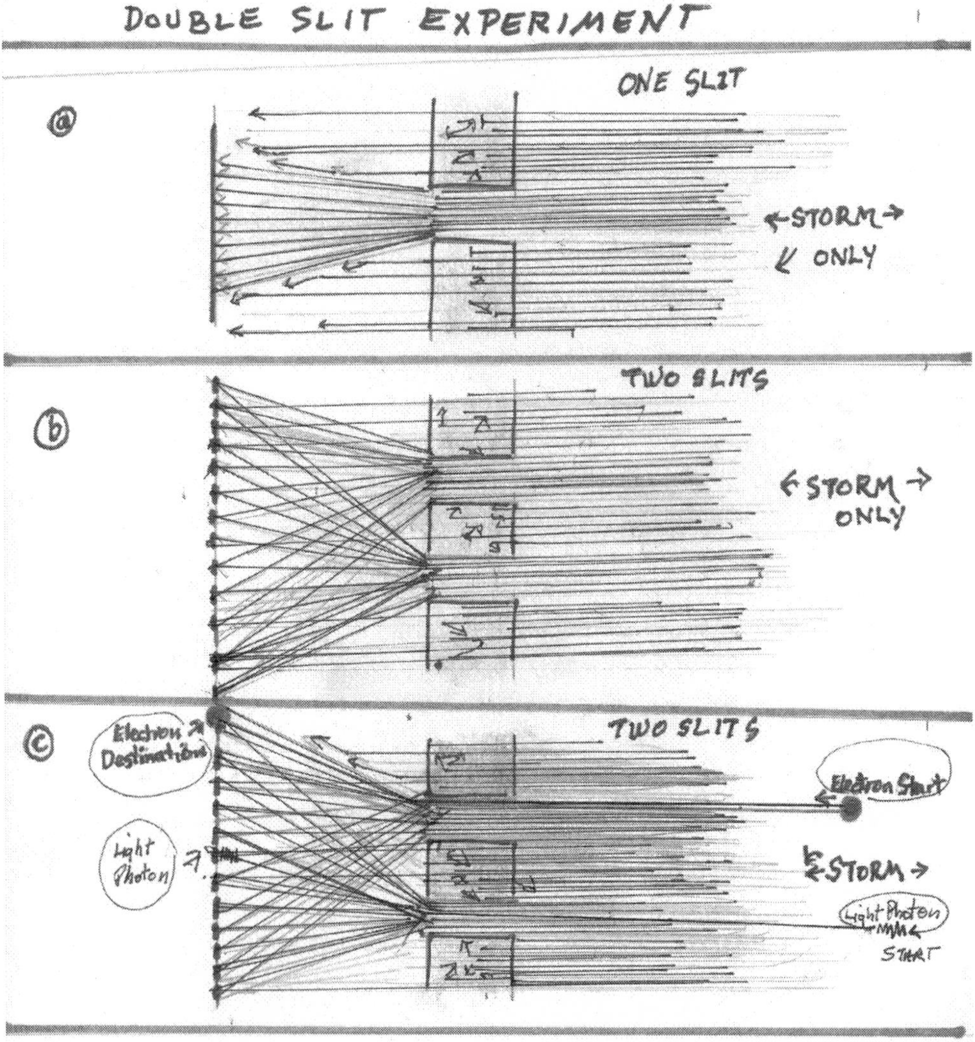

DOUBLE SLIT EXPERIMENT

ONE SLIT

ⓐ

←STORM→
ↆ ONLY

TWO SLITS

ⓑ

←STORM→
ONLY

ⓒ

Electron &
Destination

Light
Photon

TWO SLITS

Electron Start

←STORM→

Light Photon
START

See in picture a, the photon storm passes through one slit. Since fewer photons are absorbed going through the slit than going through the barrier, the photons going through the slit are more dense, so they must spread slightly to even the density around them, making up for the photons reflected by the barrier. The picture only shows the activity from one direction, but it is equally true in all directions.

In picture b, the photon-storm passes through two slits, and like the classical slit experiment, the waves interfere with each other to create alternating slats. BUT THIS PATTERN IS TOTALLY INVISIBLE! OUR EYES DO NOT SEE THE PHOTON STORM IN THE SAME WAY WE SEE LIGHT! WE SEE NOTHING THERE.

Finally, in picture c, visible light photons or electrons are sent through the slits. These individual wave/particles are not interfering with themselves; on the contrary, <u>the photon-storm is guiding them to align with the interference pattern already created by the storm.</u>

In the PS universe, no enigma exists. If we recognize the immense power of the storm, the problem is easy to visualize. We do not require mentally impossible constructions to explain the phenomena.

In the Photon-Storm universe, Einstein would BE RIGHT!

And if the double slit experiment has a simple explanation easily seen, then many other phenomena in the PS universe—deemed too complex for the human mind in the quantum universe--may also be easily visualized.

In particular, quantum entanglements in the photon storm universe have logical explanations. Elemental particles in the photon storm universe are fuzzy, but not nearly as indeterminate as in the quantum universe. Their attributes occur without any regard to the observer; these attributes are simply a product of the *stability concept* previously discussed.

In regard to "spooky" actions at distance, the configurations of god-photons allow them quite substantial lengths. Elemental particles composed of god-photons may be physically connected over long distances both by extreme lengths and by their photon storm interactions. Be aware: all particles of substance are re-created each instant by their ongoing reactions to the photon storm. The photon storm creates and controls all movements of normal photons and regular particles.

FASTER THAN LIGHT

And the speed of light can be toppled in different ways. If we pull on a long indivisible stick, the far end of the stick will move at the exact same instant as the end we pull. Have we conquered the speed of light? No, we have simply pulled on a stick. Assuming god-photons do not stretch—not necessarily true—the same light-speed-breaking phenomena may occur.

And finally—though this concept brings up a great deal of speculation—the god-photon may not be the most elemental particle in the universe. If, as previously considered, the god-photon is a product of activities on an even smaller scale, an entire universe of sub-sub-photonic particles may be behind photon construction and movement. These particles *might not* be constrained by the shared light-speed wave function as photons.

But such speculation is for future ruminations on sleepless nights. Let it suffice--the photon storm energy field puts an entirely new character into the drama of "spooky" actions and Einstein would be please.

GLOSSARY: SOME DEFINITIONS.......

THE BEGINNING: WITHOUT ENERGY, NOTHING WOULD BE.

ENERGY IS MOVEMENT: Pre-photon energy may configure in energy balls, vibrating sticks, vibrating springs, strings—somehow it re-configures with directionality to become something like a photon with directionality. We call this original photon a god-photon, with no disrespect, but only since it is present everywhere.

THE GOD-PHOTON: This smallest and most common photon is designated a god-photon because its presence is felt everywhere in all directions. More precisely, the god-photon is the smallest unit of energy that moves directionally, a quantum. Like other photons it possesses three aspects—directional movement, energy, and spin (+ -). A god-photon possesses the relative energy of Planck's constant.

The diameter of a photon must approach absolute zero, so it can pack together in incredible numbers.

THE PHOTON STORM: The photon storm is composed of tiny god-photons that move in every direction with great force. The energy of a single god-photon is miniscule, but these photons move side-by-side in such numbers, the total force in immense. The force of the photon storm is called the (S) force.

The **strong force** of the storm has the power to create matter and keep it together. The (S) force is also the force behind gravity.

THE S FORCE: The energy of the photon storm moving in all directions is very large—approximately $(1.493) (10)^{28}$ Kg/cubic meter. This is an extraordinary force. The (S) force is the **strong force** that pushes particles together. The weak force is spin in the PS model.

Questions: Does the (S) force self confine? Is this storm force presently growing or diminishing? Was the energy in our universe billions of years ago more condensed than it is now? Is the S force the same throughout the universe?

THE STRONG FORCE: The strong force in the classic model is the force that pushes elemental particles together, creating all matter. In the PS model the strong force is the photon storm or the (S) force.

THE WEAK FORCE: In the classic model, the weak force causes atoms to split and decay. In the photon storm model, the weak force is simply spin. The more spinning parts in a small area, the greater the chance of collisions. These collisions force particles to decay.

   Because of spin, the upper sizes of particle combinations are probably capped. Quarks above three or five electrons seem uncommon, and very large nucleuses are generally unstable.

THE COSMIC CONSTANT: THE COSMIC CONSTANT IS A FORCE ENVISIONED BY EINSTEIN THAT EXERTS ITS EFFECTS EQUALLY IN ALL DIRECTIONS TO CONTROL THE EXPANSION OR DIMINISHMENT OF THE UNIVERSE. THE PHOTON STORM POSSESSES THE SAME ASPECTS AS EINSTEIN'S COSMIC CONSTANT.

PHANTOM PARTICLES -- THE W, Z bosons, Higgs. The W-particle is a real particle in the PS Model, most notably seen in the decay of the neutron. The W- boson in this model is actually the debris created when two down quarks violently interact. The W- bosom exists for only microseconds until the nucleus reforms into a proton, an electron, and an anti-neutrino. See drawing page 58.

PHOTON DIRECTIONALITY: (Why do photons move?) In the photon storm model, a photon's combined wave and spin movement physically pushes it forward through a thick sea of fellow photons. The interactions of two photons, moving opposite directions, serves to aid the movement forward of both photons. In this model, photons cannot move through an absolute vacuum. Spin, clockwise and counterclockwise, assists photons to easily move over or under each other.

   In the classic model, the reason for photon movement is unknown.

LIGHT SPEED: In the PS model, since all photon movement is conjoined, photons move at the same speed, regardless of their origination. Moving through materials, such as people or glass, large photon strings refract more than smaller ones, due to the respective sizes of their photon numbers. In the classic model, photon number is called frequency. In the classic model, light speed is constant regardless of origination, as it is in the photon storm model, but the classic model presents no physical reason for the constant speed of light.

PLANCK'S CONSTANT (or h): The minimal difference between the energy of photons or photon bundles of different strength. The energy of photons does not increase linearly but in incremental steps. Planck's constant is a measurement of **one** of these steps.

Values of h	Units	Ref.
$6.626070040(81) \times 10^{-34}$	J·s	[1]
$4.135667662(25) \times 10^{-15}$	eV·s	[2]
2π	$E_P \cdot t_P$	
Values of $\hbar$ (h-bar)	**Units**	**Ref.**
$1.054571800(13) \times 10^{-34}$	J·s	[2]
$6.582119514(40) \times 10^{-16}$	eV·s	[2]

COMBINATION PHOTONS OR **PHOTON THREADS:** IN THE PS MODEL ALL LARGE PHOTONS may simply be COMBINATIONS OF SMALLER GOD-PHOTONS. PHOTON THREADS HAVE THE SAME ASPECTS AS THE ORIGINAL GOD-PHOTON, ENERGY AND SPIN. THE GREATER NUMBER OF GOD-PHOTONS THREADED TOGETHER IN A COMBINATION PHOTON GIVES COMBINATION PHOTONS MORE ENERGY THAN A SINGLE GOD-PHOTON. OTHERWISE GOD-PHOTONS AND REGULAR PHOTONS MOVE SIMILARLY.

Light energy is an example of combination photons, as are micro wave energy, infrared energy, and radio waves, plus extraordinarily powerful gamma rays from space.

Frequency: The number of photons in a threaded bundle corresponds to measured frequency in the classic photon.

Amplitude: The number of separate photon bundles measured corresponds to amplitude. (Although some high school teachers still teach that individual photons possess amplitude, most physicists agree even the classic wave photon does not possess amplitude. Amplitude, again, is a measure of photon number.)

Energy: Every god-photon possesses the same energy, approximately one Planck. The energy of a thread bundle of photons equals the number of god-photons in that bundle times Planck's constant. This is the same as (frequency times Planck's constant equals energy), the formula in the classic model.

CLASSIC PHOTON: In the classic quantum model, photons are both particle bundles and waves. These waves possess different frequencies according to their energy levels. The greater the frequency of a photon, the smaller is the wavelength of that photon, or distance between wave tops, and the greater is the energy of that photon.

So powerful gamma rays have very short wavelengths and high frequencies. Light waves we can see have longer wavelengths. Radio waves, which move through our bodies without much interaction, have longer wavelengths still and, of course, lower frequencies and lower relative energy. In general, the penetrating ability of photon waves increases as their wavelengths increase and their energies decrease.

The energy of classic photons does not increase linearly, but in tiny increments, measured as Planck's constant.

The classic photon model does not provide a reason why photons move. The classic photon model does not provide a visual model for how photons are produced or absorbed by matter.

MASS: Mass is a measure of the inertia of an object. Stated differently, mass is an object's resistance to movement. The more massive an object is, the more difficult it is to move it. Mass or inertia in the PS universe is caused when the photon storm pushes on an object from all directions. Without such a push, an object would have no mass or inertia. Indeed, without the push from the photon storm objects would have no shapes whatever and would change back into the many photons from which they are created.

Photons have no inertia or mass simply because the photon storm does not push on other photons. Photons move through the storm without friction interaction.

In the PS model: Mass = (Rf) (Cubic area) (S). Or the mass of an object equals its reflectivity times its cubic area time (S) or the storm force.

CLASSIC GRAVITY: In Newton's classic universe, every particle of mass pulled invisibly on other mass. This pull was gravity.

Einstein altered Newton's concept. Mass warps or condenses space-time; this warping or condensing of space-time pulls objects in space closer.

Newton's gravity calculations suffice for normal objects. Einstein's calculations are more accurate for very large objects and speeds approaching that of light.

PS GRAVITY: In the PS universe, the photon storm pushes all objects with mass together. This gravity push is created because objects with mass reflect or absorb a small amount of the storm's energy. Thus the space between objects has a smaller energy push than the area outside the shadow.

Picture a large object being pummeled from every direction by the storm. The storm gives this object integrity and mass. Now place this object next to another object. Both objects reflect a tiny part of the storm, so the area between the objects is less intense than areas where no reflection occurs. This small difference in force pushes the two objects together. Gravity is a very small force when compared to the total force of the Storm, since most of the storm passes through objects without interaction.

Gravity calculations in the PS universe conform to Einstein's calculations. Whether space-time is warped to create gravity, or whether space defined by energy creates mass and matter—the results are the same. (See math section)

REFLECTIVITY (Rf): Particles possess mass because they reflect/absorb a small amount of the storm's power--the greater a particle's reflectivity, the greater its relative mass. Cubic area times reflectivity times (S) = total mass.

Large dense objects such as stars and black holes possess greater reflectivity than normal objects such as planets and people.

Particles, such as photons, possess **no** reflectivity, so they possess **no mass**. In other words, photons do not reflect other photons, but manage to move over each other easily.

Reflectivity increases when a particle's configuration is stable and attracts additional fluff from the storm, such as in electrons and quarks. So small changes in configuration can lead to large changes in mass.

NEUTRINO: A configuration of god-photons, more stable than a photon thread, but not yet as stable as the electron configuration. The neutrino configuration attracts very little mass and has no measurable charge. Neutrinos apparently exist in massive numbers; their abundance makes sense since they are a very simple configuration closely related to god-photons.

Neutrinos apparently may be re-configured into electrons and other particles when they interact with atoms. Perhaps contact forces a neutrino into point stability, or creates some other unknown stability that becomes matter. Otherwise, neutrinos—like god-photons--pass through planets with little consequence.

ELECTRON: A very stable configuration created by god-photons. This electron stability may be created by several photons meeting at the exact same point. This **point stability** attracts mass. An electron is similar to a large photon, possessing the same attributes of directionality and spin, but also attracting mass.

Since an electron spins like a photon in clockwise and counterclockwise directions, the electron spins the god-photons around it into vortexes. These tornado-like vortexes focus the energy of the storm and exert pulls on other electrons, pulls that we measure as electric charges. These charges are plus and minus depending on the clockwise or counterclockwise spin of the electron and its associated vortexes.

Electrons create both positive and negative spins, but the charge in line with the electron's directionality is what primarily defines the electron. We have **arbitrarily** assigned a clockwise spin to a regular negative electron and a counter-clockwise spin to a positron. Thus we have regular electrons with a negative charge of (-1) and anti-electrons or positrons with a positive charge of (+1) in line with their directionalities.

POSITRON: A positron or anti-electron is an electron spinning contrarily to a regular electron. A positron has a positive directionality or a charge or + 1. When a positron electron and a negative (normal) electron collide, they can unravel into their constituent photons. This unraveling releases significant energy in the form of gamma rays.

However, combined at an angle in the same direction, positrons and electrons would have spins that combine. We suggest these form quark point configurations in the creation of larger particles, such as the proton and neutron.

CHARGE: Each electron creates a photon vortex, a tornado that pulls or pushes on other electrons depending on their spins. The force of these vortices is called charge, and is plus or minus one, depending on spin directionality. Spins moving the same direction push each other. Vortices moving opposite directions pull each other.

When many free electrons are aligned in the same directions, the combination creates a magnetic field. The vortices of many electrons combine to create a complete vortex circle extending between the positive and negative anodes of each electron.

In non-magnetized materials, electrons are arrayed randomly and are tightly bound to nucleuses, so electron charges do not combine to create force fields in any one direction. Only metallic materials that possess free electrons that point in the same direction allow charges to be combined creating magnetism.

At low temperatures, however, some non-magnetic materials can become super-conductors or super magnets when their photon flows diminish to the point electrons attach weakly.

In other words, the flow of Bourne-fluff photons through an atom—also known as heat—helps align an electron in its position in an atom. When the photon flow of fluff—or heat energy—diminishes severely, the electrons become free to act as super conductors or as super magnets.

ELECTROMAGNATISM: Electromagnetic radiation is photon radiation. If an electron is in a magnetic field, its positive and negative vortices will be pulled in opposite directions, increasing the magnetic force around the electron. However, if the magnetic field is suddenly released, the vortex will relax. A thread of spinning photons, no longer tied in to an electron, will release as new free photons.

These spun-together photons are what we observe as light and feel as infrared energy. These photons interact with plants to promote the chemical changes for growth.

These spun-together photons can be absorbed by matter and later re-emitted at different energy values—though the total energy is always conserved.

When an electron absorbs a photon bundle it increases in speed and energy. When an electron emits a photon bundle its speed and energy decrease according to Einstein's formulas.

HEAT ENERGY--PHOTON FLUFF, BOURNE-FUZZ, PHOTON BUNDLES, GOD-PHOTON NUMBERS: Heat energy is simply a measure of the number of god-photons added or subtracted from a configuration. When an electron absorbs a photon bundle it absorbs heat energy. When an electron or a nuclear package emits a photon bundle that particle loses heat energy.

At absolute zero, a particle has emitted nearly all its absorbed photon fluff energy.

GAMMA RAYS: Gamma rays are the most powerful photons twisted together in atomic reactions and stars. They move like ordinary photons but consist of mega-numbers of god-photons.

Like other photons, they possess no mass. But they are so powerful they can do damage to ordinary matter, knocking electrons out of orbits and even knocking apart molecules.

Question: Do gamma rays have upper energy limits beyond which they turn into matter or something else?

Historically, all particles with frequencies greater than about 10^{19} Hertz (or about 50,000 electron Volts (5×10^4 eV) where a typical optical photon carries 2-3 eV) **are** called **gamma-rays**. Theoretically, there is no hard **limit** to the **energy** that a **gamma-ray can have**.Apr 12, 1997

POINT STABILITY: Point stability is a **possible** stable configuration achieved when elemental particles **possibly** circle around the exact same point. **If and how** particles achieve point stability is still conjecture. Some physicists speculate that if two gamma ray photons collide directly, they may create matter in the form of electrons. Gamma rays may possess the power necessary to squash photons into some sort of stable configuration called matter. Photons in such a mash may be point stable or they may be wadded in some other way to keep them together. Point stability, however, seems the most *graceful* way to glue things together.

Neutrinos may be particles not yet point stable. When neutrinos encounter atoms directly they may be re-configured into point stable electrons or muons.

Random encounters between god-photons **might** occur to create point stability also—and these encounters might require very little energy--but such encounters would be rare, bearing in mind a photon's ability to avoid contact as it moves through space. But space is rather large, so such encounters are still possible. **And if**, around the time of the big bag, space happened to be more condensed, these encounters might be more likely.

If such encounters actually occur, given enough time, electrons could be created directly from the photon storm.

If a photon storm universe possessed significant amounts of electrons, then these electrons would gravitate together, and meeting together closely, new alignments for quarks, protons, and neutrons might also be created. The condensed state of a big bang would facilitate these creations.

Of course, the entire scenario is conjecture, and it may be too simplistic to be reality.

QUARKS: Quarks are electron combinations that have achieved point stability. An **up quark** is the simplest point combination of two positron/antielectrons and a single normal electron. An up quark has a vector charge of + 2/3. A **down quark** consists of three regular electrons and two positron/antielectrons. A down quark has a vector charge of -1/3.

QUARK MASS: Electrons composing quarks gather **fluff** from the storm that increases their mass approximately 140 to 170 times their original mass. Fluff forms on elemental particles in shells, rather like water crystals on snowflakes.

FLUFF, BOURNE FLUFF, FUZZ, PHOTON ENERGY*: In the PS model fluff accounts for most of the mass in the universe.* Fluff apparently accumulates on all configurations that are not pure photons. Photons move effortlessly through the cloud, so, no matter how large they become, the storm reacts to them exactly as another photon. They are massless. Photons attract no fluff.

However, something that is not a photon, in other words, moves slower than the speed of light, or is more awkward than a photon, clashes or bounces against all the photons constantly flowing over and moving around it.

Thus an electron may be composed of photons and possess many photon characteristics, but it does not move effortlessly through the storm. An electron's configuration creates inertia also called mass.

All other electron re-configurations, such a quarks, also attract mass. The amounts of mass or Bourne-fluff these configurations attract depend on their combined configuration. A proton configuration that may consist of only eleven electrons weighs the same as 1835 electrons. The difference is caused by fluff.

Thus very simple configurations may produce massive particles. On the contrary, vast strings of god-photons threaded in gamma rays will possess no mass but great energy.

FLUFF FLOW: Fluff flows constantly through and around the configurations of elemental particles. Fluff flow and charge help keep larger particles together. Imagine streams of energy flowing through every part of atoms and molecules, being emitted and replenished constantly by the photon storm. This flow can be measured as heat.

PROTON: A PROTON CONSISTS OF TWO UP QUARKS AND ONE DOWN QUARK COMBINED AT A SINGLE POINT, WITH SPINS MESHING. THE PROTON HAS A VECTOR CHARGE OF + 1.

NEUTRON: A NEUTRON CONSISTS OF TWO DOWN QUARKS AND ONE UP QUARK AND HAS A NEUTRAL CHARGE OF 0.

NEUTRON DECAY: A NEUTRON CAN DECAY INTO A PROTON WHEN ITS TWO DOWN QUARK ARMS CRASH INTO EACH OTHER.

HADRONS: Particles created by the combinations of quarks.

LARGER PARTICLES: PARTICLES LARGER THAN A NEUTRON ARE POSSIBLE, BUT THEY WOULD APPARENTLY BE LESS STABLE. INCREASING SIZE MAKES THE POSSIBILITIES OF QUARK CLASHES INCREASINGLY PROBABLE.

ANTI-PARTICLES: IN THE PS MODEL ANTI-PARTICLES SPIN THE OPPOSITE DIRECTION FROM THEIR PARTICLE TWIN. THE POSSIBILITY OF ANTI- PARTICLES IS EQUAL TO THE POSSIBILITY OF PARTICLES, SO ANTI-PARTICLES SURELY EXIST. WHEN A PARTICLE MEETS ITS ANTI PARTICLE HEAD-ON, THE TWO PARTICLES MAY UNRAVEL INTO THEIR ORIGINAL PHOTONS. THE UNRAVELING RELEASES A GREAT DEAL OF ENERGY.
   HOWEVER, PARTICLES AND ANTI-PARTICLES MAY JOIN TOGETHER THEIR SPINS AT A POINT AND BECOME STABLE QUARKS, PROTONS, AND NEUTRONS.

PROTON NEUTRON PAIR: A NEUTRON AND A PROTON HELD TOGETHER ARE MORE STABLE THAN OTHERWISE. THE BOND CIRCULATES PHOTON ENERGY THROUGHOUT THE COMBINATION AND HOLD THE NEUTRON ARMS APART.

ALPHA PARTICLE: AN ALPHA PARTICLE IS A NAKED HELIUM NUCLEUS CONSISTING OF TWO PROTONS AND TWO NEUTRONS. AN ALPHA PARTICLE IS RELEASED FROM MANY NUCLEAR REACTIONS.

ELECTRON ORBITS: Electrons do not orbit nucleuses as is sometimes still taught in schools. Electrons flow around nucleuses in ways that accommodate contact with the configurations of the core.

BETA PARTICLE: A BETA PARTICLE IS A HIGHLY ENERGIZED ELECTRON, ALSO OFTEN RELEASED FROM NUCLEAR REACTIONS.

A GAMMA RAY: A GAMMA RAY IS AN EXTREMELY LARGE PHOTON THREAD, RELEASED FROM NUCLEAR REACTIONS ON EARTH AND FROM STARS. LIKE OTHER PHOTONS IT HAS NO MASS, BUT IT HAS DANGEROUS POWER AND CAN KNOCK ELECTRONS FROM THEIR POSITIONS AND DISTURB MOLECULAR RELATIONSHIPS.

THE POSSIBILITY THAT GAMMA RAYS CAN BE SMASHED TOGETHER TO CREATE ELECTRONS IS A CURRENT TOPIC OF PHYSICS. WHEN AN ELECTRON AND POSITRON or ANTI-ELECTRON COLLIDE, THE RESULTS ARE GAMMA RAYS. COLLIDING GAMMA RAYS SHOULD THEREFORE PRODUCE ELECTRONS AND ANTI-ELECTRONS.

CREATING MATTER FROM GOD-PHOTONS: IF THE POINT CONFIGURATION THEORY IS CORRECT, A LESS ENERGETIC WAY TO CREATE MATTER WOULD BE TO FOCUS GOD PHOTONS AT APPROPRIATE ANGLES TO CONFIGURE TOGETHER. FOCUSING NEUTRINOS MIGHT ALSO BE A POSSIBLE METHOD TO CREATE MATTER. NEUTRINOS CIRCULATE IN VAST NUMBERS.

IF SUCH A FINESSE TECHNOLOGY WERE PROVEN POSSIBLE, A GREAT DEAL OF ENERGY IN THE FORM OF MATTER COULD BE CREATED FROM A SMALL AMOUNT OF ORIGINAL ENERGY. IF ELECTRONS AND POSITRONS WERE CREATED, THEY COULD LATER BE TURNED BACK INTO ENERGY, USEFUL FOR FUTURE CIVILIZATIONS. BUT "IF" IS A BIG WORD.

THE EDGE OF THE UNIVERSE: AN INTERESTING QUESTION REVOLVES AROUND WHAT MIGHT BE OUTSIDE THE PHOTON STORM UNIVERSE. IN THE CLASSIC UNIVERSE, A GREAT VOID/VACCUUM MAY STRETCH TO INFINITY. IN A PHOTON STORM UNIVERSE, MATTER CANNOT EXIST IN A COMPLETE VACUUM. SO OUTSIDE THE PHOTON STORM UNIVERSE, NOTHING—NOT EVEN A VACUUM--MAY EXIST. ASSUMING A PHOTON STORM UNIVERSE SELF-CONFINES ITSELF, THEN AN INFINIT VOID MAY SIMPLY NOT BE THERE. THE INFINIT VOID MAY NOT

EXIST, SO WE DON'T HAVE TO WORRY ABOUT IT. IT IS ONLY A FIGMENT OF IMAGINATION?? HM?

THE PS UNIVERSE: If the photon storm dominates the universe, then Einstein's Cosmic Constant Universe is also possible. Imagine a huge ball filled with god-photon energy. At the edge of this universe, most, if not all of the photons turn around and move back towards the middle.

If the turnaround scenario is true, the entire ball could become somewhat stable, a vast ball of photon storm energy. This energy may be more condensed at the core of the ball than at the edges, since more directional energy is directed to the center than to the edge.

So matter created in any way would have a slight push to the outer edge of the ball. Otherwise, this huge ball would enjoy a reasonable constancy. It would be COSMICALLY CONSTANT as Einstein envisioned.

Variations from its constancy might occur if the big ball bounces against other big or small balls of various possibilities at its edges; and, of course, an odd explosion of a nearby mega black hole might add some drama.

But otherwise, the configuration could enjoy some stability and be as Einstein envisioned, COSMICALLY CONSTANT.

THE DOUBLE-SLIT EXPERIMENT: The double slit experiment, first performed in 1801, demonstrates the double nature of light photons. When photon light moves through a single slit the light refracts beyond the slit and forms a wide refraction pattern. However, when light moves through two slits, these refraction patterns interfere creating patterns of light and dark. The assumption is, light is a wave front, and thus the addition or subtraction of the light waves, depending on their wavelengths, causes the interference patter. Later, scientists discerned that light impacted at specific points. The conclusion was that light acted both as a wave and as a distinct particle. Even later, to everyone's surprise, when single photons or electrons were sent through slits, these particles apparently continued to interfere with each other as a wave pattern. The final conclusion was that single elemental particles were able to interfere with themselves by taking alternate routes through time. Though the conclusion was mind bending, probability mathematics seemed to verify the results.

De Broglie–Bohm theory[edit]

An alternative to the standard understanding of quantum mechanics, De Broglie–Bohm theory states that particles have precise locations at all times, and that their velocities are influenced by the wave-function. So while a single particle will travel through one particular slit in the double-slit experiment, the so-called "pilot wave" that influences it will travel through both. The two slit de Broglie-Bohm trajectories were first calculated by Chris Dewdney whilst working with Chris Philippidis and Basil Hiley at Birkbeck College (London)[63]. The de Broglie-Bohm theory produces the same statistical results as standard quantum mechanics, but dispenses with many of its conceptual difficulties.[64]

THE DOUBLE-SLIT experiment in the PHOTON STORM MODEL: In the photon storm model, the double slit experiment is easier to understand, since it takes place in a storm of photons with vast power. The power of the photon storm acts as a "pilot wave" similar to that of the De Broglie-Bohm theory. The storm forms its own interference pattern invisible to observers, and this pattern guides elemental particles that also travel through the slits. So a particle does not need to interfere with itself.

THE BIG BANG: The entire universe may have originated from a tiny point around thirteen billion years ago, from a big bang expansion. The reason astronomers believe the big bang occurred is that they can look deeply into the universe and view light that originated billions of years ago that only reaches us now. They are looking back in time. This look back in time tells astronomers the universe is apparently expanding. When astronomers look at the edge of the universe, they perceive the expansion is apparently increasing. A chatter of very low energy electromagnetism emanates from every direction, a leftover, they think, from the energy of the original big bang.

If the big bang theory is correct, the universe should continue to expand indefinitely, unless it encounters an opposing force. Gravity is not strong enough to interfere with the expansion.

SPACE ELASTICITY: "The universe abhors a vacuum," is a statement widely quoted in thermodynamics. The concept of space elasticity derives from the concept, "Perhaps a vacuum abhors itself." In other words, perhaps a vacuum dislikes being a vacuum so much it resists existing. To add to the confusion, empty space *may* possess a counter-force to energy expansion.

In the unlikely event this is true, a certain amount of energy would be necessary to define any area of space, like air blowing up a balloon.

Space elasticity—if it exists--would help keep a ball of energy, like a universe, together. But such elasticity is not absolutely necessary if energy possess a tendency to self-confine.

THE STABILITY CONCEPT: The stability concept (also called the concept of the obvious) merely states that what is stable will outlast what is unstable over time.   Elemental particles stable over eons will eventually dominate over particles stable for microseconds.

So our universe is composed of the particles that demonstrate stability.

In biological terms the stability concept morphs into *Darwin's survival of the fittest*. The most stable life strategies succeed over strategies that are less stable. If we look at life as a single unity, over several billion years life has out-survived continents and mountains. Intelligent life may outlast our solar system and even our universe. If then, the universe is a test of stability, life may be *it*. Life *may* be the most stable of all things. God may have known what She was doing.

THE STABILITY CONCEPT VS THE LAWS OF THERMODYNAMICS: THE STABILITY CONCEPT DOES NOT CONTRADICT THE LAWS OF THERMODYNAMICS. HOWEVER, STABILIZING FACTORS ACT IN OPPOSITION TO THERMODYNAMIC FORCES OF ENTROPY. ONE DAY THE UNIVERSE MAY EXPAND TO INFINITY OR CONTRACT TO A SINGLE POINT. IF WE WAIT LONG ENOUGH, WE MAY KNOW WHICH FORCE HOLDS THE UPPER HAND.   BUT WE SHOULDN'T HOLD OUR BREATHS.

Made in the USA
Monee, IL
19 August 2021